Memetic War

Memetic War analyses memetic warfare included in cyber war and aims to develop a framework for understanding the parameters included in utilising this concept in Ukraine as a part of civic resistance.

In the Ukrainian war, an informal defence tactic has developed to uphold the information flow about the war and to debunk Russia's communications. The war has enhanced the visibility of governmental and civic activation by using the advantages of social media architecture, networks, and communication forms. The book investigates Ukraine's public and private abilities to develop cyber capabilities to counter propaganda and dis-and-misinformation online as a defence mechanism. This book uses military ROC doctrine to understand government authorities, the armed forces, and civic engagement in the Ukrainian resistance.

Memetic War will have relevance for scholars, researchers, and academics in the cybersecurity field, practitioners, governmental actors, and military and strategic personnel.

Tine Munk is a Senior Lecturer in the Criminology and Criminal Justice Department at Nottingham Trent University, UK. Tine predominantly teaches and researches cybercrime and cybersecurity. Her overarching research interest is cybercrimes in a political context, focusing on these crimes' power, responses, and impacts.

Routledge Studies in Crime and Society

For more information about this series, please visit: www.routledge.com/
Routledge-Studies-in-Crime-and-Society/book-series/RSCS

Memetic War

Online Resistance in Ukraine

Tine Munk

Routledge
Taylor & Francis Group

LONDON AND NEW YORK

First published 2024
by Routledge
4 Park Square, Milton Park, Abingdon, Oxon OX14 4RN

and by Routledge
605 Third Avenue, New York, NY 10158

Routledge is an imprint of the Taylor & Francis Group, an informa business

© 2024 Tine Munk

British Library Cataloguing-in-Publication Data
A catalogue record for this book is available from the British Library

Library of Congress Cataloging-in-Publication Data
Names: Munk, Tine, 1970– author.
Title: Memetic war : online resistance in Ukraine / Tine Munk.
Other titles: Online resistance in Ukraine
Description: Abingdon, Oxon ; New York, NY : Routledge, [2024] |
Series: Routledge studies in crime and society | Includes bibliographical
references and index.
Identifiers: LCCN 2023022060 (print) | LCCN 2023022061 (ebook) |
ISBN 9781032558639 (hbk) | ISBN 9781032558653 (pbk) |
ISBN 9781003432630 (ebk)
Subjects: LCSH: Information warfare—Ukraine. | Cyberspace operations
(Military science)—Ukraine. | Information warfare—Russia (Federation) |
Memes—Ukraine. | Civil-military relations—Ukraine. | Ukraine—
Foreign relations—Russia (Federation) | Russia (Federation)—Foreign
relations—Ukraine.
Classification: LCC U167.5.C92 M86 2024 (print) |
LCC U167.5.C92 (ebook) | DDC 355.4/7477—dc23/eng/20230627
LC record available at https://lccn.loc.gov/2023022060
LC ebook record available at https://lccn.loc.gov/2023022061

ISBN: 978-1-032-55863-9 (hbk)
ISBN: 978-1-032-55865-3 (pbk)
ISBN: 978-1-003-43263-0 (ebk)

DOI: 10.4324/9781003432630

Typeset in Times New Roman
by codeMantra

Inge Højsgaard Munk
22.11.1942–06.01.2018

Contents

Figures

Tables

Boxes

Abbreviations

DDoS – Distributed-Denial-of-Service
DPR – Donetsk People's Republic
EU – European Union
ISIS – The Islamic State of Iraq and the Levant Islamic State of Iraq and Syria
LPR – Luhansk People's Republic
MAP – Membership Action Plans
NATO – North Atlantic Treaty Organization
NAFO – North Atlantic Fella Organisation
NGOs – Non-governmental Actors
ROC – Resistance Operating Concept
UMF – Ukrainian Meme Force
UN – United Nations
US – United States of America
USSR – Union of Soviet Socialist Republics
OSCE – Organization for Security and Co-operation in Europe

Acknowledgements

This book owes its existence to the invaluable support and encouragement of a diverse group of people who supported me from the sidelines while I was researching topics such as Ukraine, war and warfare, communications, symbols and memes, behaviours and language, resilience and resistance. The journey was challenging and emotional, and I am grateful for the support and assistance I received from friends, family, colleagues, and students. I want to express my appreciation to those who played a role and acknowledge this community. This research journey began on 24 February 2022, when I started researching news about Ukraine and online warfare, but it quickly turned into systematic online ethnographic research. The book resulted from extensive writing during the festive break of 2022–2023. Having followed war and warfare very closely, I want to extend my special gratitude to everyone fighting for freedom and justice, for themselves or others, online or offline.

It is heartening to have such a large group of supporters who contributed to this book in various ways. I want to thank Medha Malaviya at Routledge, for her patience, support, and belief in this project. I would also like to express my gratitude to the editorial board for considering this book timely and relevant to a broad audience; to Sarahjayne Smith at Routledge for her work on the production side; and to Helen Baxter from HJB Editorial for copy editing the manuscript.

Special thanks go to "Blå Gruppe" (as always), Maibritt and Jan Falkesgaard, Inge and John Ankjær Bertz, Laura and Jesper Munk – and Generation 2.0, Marcus and Sarah Ankjær Betz, and Emma, Emil, and Jeppe Munk. I would also like to thank Evald Munk and Ruth Rousing.

During this research, I received much support from a large group of people who played a crucial role in conducting an intense, challenging, and heartbreaking study of war, its devastation, and the gross human suffering that followed. First and foremost, I want to thank Elliot Doornbos for his enthusiasm in discussing this book and my research. I also express my deep gratitude to my cyber colleagues at Nottingham Trent University for creating an excellent and supportive environment that allows me to explore edgy cyber interests. Thank you, Thais Sarda, Martin Tangen, and Philip Wane. Special thanks go to my colleagues and friends who supported this project unconditionally and excitedly from the start, including Tine Lee Senft, Bettina Jacobsen, Lene Dam, Magali Peyrefitte, Graham Smith, Becs Waterman, Georgia Mouskou, Sara Rodriguez, Angus Nurse, and Vicky Kemp.

I could not research and write without constant communication with students to test ideas and perspectives. I want to thank my students at Nottingham Trent University, specifically the Criminology and Criminal Justice Department's third-year criminology students in 2022–2023. They listened and engaged in my first ever lecture about memetic warfare in January 2023. My PGR student, Juan Ahmad, has also been patient with me and my constant talk about the memetic war. I am grateful for his willingness to engage in discussions about this topic.

1 Introduction

Tine Munk

Introduction

2022

As 2022 started on the ruins of the Covid-19 pandemic, not many Europeans could have predicted that the next catastrophe would be a war on the doorsteps of Europe. It was a common belief that counties would begin to recover economically after the pandemic. Vaccine programmes were developed, and citizens lived with the virus; there was no appetite for conflicts or wars. Nevertheless, Russia launched a full-scale invasion of Ukraine without any provocation in advance. The timing of the invasion happened when United States' (US) President Biden's administration demonstrated a reluctance to be involved in prolonged military conflicts, and the presidency was dealing with internal troubles in the aftermath of President Trump's election loss in 2020, and the attack on Capitol Hill on 6 January 2021 (BBC News, 2022a; Donovan et al., 2022, pp. 314–319; Harris et al., 2022). The invasion also occurred in the middle of profound political, economic, and societal changes in Europe. The Western bloc was polarised, and states had enough internal issues to deal with. Europe was still struggling with the Covid-19 pandemic. Political changes loomed in Germany after Chancellor Merkel left office. French President Macron faced a re-election campaign against a resurgent right-wing candidate, and the United Kingdom (UK) struggled with the post-Brexit economy and political chaos. Russia had long relied on Europe's dependency on Russian oil and gas, so the Kremlin did not expect severe reactions beyond what they experienced in 2014 after the annexation of Crimea and the war in Donetsk and Luhansk. President Putin and the Kremlin apparatus seized the moment to escalate the Ukrainian war while the Western world was otherwise occupied (Davidson Sorkin, 2021; David, 2022; Harris et al., 2022; Leicester, 2022; WEF, 2022; Ziady, 2022;).

Snowballing the "Special Military Operation"

In 2022, Russia pursued an endgame to the military campaign that had begun in 2014. On 21 February 2022, Russia recognised two Ukrainian "rebel

DOI: 10.4324/9781003432630-1

regions", Donetsk and Luhansk (also known as the Donbas region), as independent states to pave the way for the invasion. Russia argued that the two new independent regions asked the country to enter part of Ukraine on an artificial peacekeeping mission (CRS, 2022, p. 2; United Nations, 2022). Globally, this was the first step for the more significant and extensive invasion of Ukraine. US President Biden and the United Nations' (UN) Secretary-General Guterres quickly condemned the strategic move to assist the two "independent" states.

However, Russia ignored these condemnations and launched a full-scale invasion of Ukraine on 24 February 2022. Russian troops entered Ukraine from several directions, including Crimea, the Donetsk and Luhansk regions, and the border with Belarus. The invasion was swift and brutal, with Russian forces quickly seizing control of several Ukrainian cities and key infrastructure. The conflict rapidly escalated into an all-out war, with heavy casualties and significant damage to infrastructure (Munk & Ahmad, 2022; O'Connor, 2022). The international community, including the UN, the European Union (EU), and many individual countries, have widely condemned the actions taken by Russia in Ukraine. The invasion has resulted in significant loss of life and displacement of civilians and has had far-reaching economic and political consequences. The North Atlantic Treaty Organization's (NATO) Secretary-General Stoltenberg reacted swiftly to the invasion by condemning the action. Stoltenberg argued that this was a reckless attack initiated by Russia that, in spite of an intense period of international diplomacy, had decided to follow the "path

Table 1.1 US President Biden's Reaction to President Putin's Announcement of the Two Independent States

Speaker	Quote
US President Biden, 21 February 2022	Who in the Lord's name does Putin think gives him the right to declare new so-called countries on territory that belonged to his neighbours? This is a flagrant violation of international law, and it demands a firm response from the international community

Source: (Liptak, 2022; The White House, 2022; Smith, 2022).

Table 1.2 UN Secretary-General Guterres' Reaction to the Russian Self-proclaimed Peacekeeping Mission

Speaker	Quote
The UN Secretary-General Guterres, 23 February 2022	[W]e meet in the face of the most serious global peace and security crisis in recent years – and certainly in my time as Secretary-General. Our world is facing a moment of peril. I truly hoped it would not come

Source: (United Nations, 2022; UNIS, 2022).

of aggression" (NATO, 2022; O'Connor, 2022). The European Commission President, Von der Leyen, echoed this argument by claiming that President Putin was: "[R]esponsible for bringing war back to Europe" (BBC News, 2022b; O'Connor, 2022).

The Russian Invasion, 24 February 2022

Russia launched a "special military operation" [full-scale invasion] to "demilitarise and de-Nazify" Ukraine. Russia's justification for the invasion has been disputed by many, and the evidence put forward to support their claims has been called into question. The accusation that Ukraine posed a threat to Russian interests and ethnic groups has been viewed by many as unfounded, and the invasion has been seen as an attempt to expand Russian influence and control (see Chapter 3) (CRS, 2022, p. 6; Fisher, 2022; Kirby, 2022; Lister et al., 2022;). Russia also argued that a potential Ukrainian NATO membership would threaten Russian sovereignty and interests (see Chapter 2). Yet, the invasion was a significant strategic miscalculation by the Russian President and his Kremlin inner circle, as they believed they would be received as the liberators and the Ukrainians would welcome them with flowers and cheers (Harding, 2022). The response of the Ukrainian people and armed forces has been praised for its resilience and determination. The conflict has highlighted the strength of Ukrainian nationalism, the desire for freedom and self-determination, and the risks and challenges posed by foreign intervention and aggression. (CRS, 2022, p. 2; Harari, 2022).

The Full-scale War

The original strategy was seizing Kyiv within a few days (Barrington, 2022; Epstein & David, 2022; Lutska, 2022; Marson, 2022; Munk & Ahmad, 2022). However, the Kremlin authorities failed to consider that Ukraine had mobilised and reformed its armed force since 2014 when Russia illegally annexed Crimea and started the war in the Donetsk and Luhansk regions. Ukraine's military doctrine and structure changed to incorporate NATO's principles and standards (Economist, 2022; Herszenhorn & McLeary, 2022).

In the early days of the invasion, the Russian armed forces surrounded the Ukrainian capital, Kyiv, and attacked the country's southern, eastern, and northern regions. The Russian armed forces made progress in all areas, but most significantly in the south. The armed forces of Ukraine took control of the areas around Kyiv in mid-March/beginning of April. The successful Ukrainian counterattacks forced Russia to abandon its attempt to take the capital. Instead, a decision was made to regroup troops in the south, east, and northeast of the country, where the Russian military had aready seized large areas (BBC News, 2022; CRS, 2022, pp. 7–8, 11; Global Conflict Tracker, 2022; O'Brian, 2022).

The Russian army continued its offensive by encircling major urban centres such as Sumy, Kharkiv, and Chernihiv (CRS, 2022, pp. 5, 11). However, by September 2022, Ukraine launched a large-scale, well-coordinated counteroffensive in the northeast, pushing back the Russian forces. The Ukrainian counter-operation opened a large front and encircled the Russian military, forcing troops to withdraw from the area (BBC News, 2022; Global Conflict Tracker, 2022; Sabbagh, 2022). This counteroffensive prompted Russia to officially annex four oblasts – Kherson, Zaporizhzhia, Luhansk, and Donetsk – that were partly occupied by Russia. After a fake referendum in September orchestrated by the Kremlin authorities, President Putin signed the accession treaties (CBS News, 2022; Sauer & Harding, 2022).

The advance of the Ukrainian armed forces was successful in liberating 3,000 square kilometres in just six days. The operation caused a disintegration of Russian forces around logistically important cities, such as the Kherson and Kharkiv offensives. The counter-operation was carefully planned, focusing on military build-up, operational security, and an effective military strategy (BBC News, 2022c; CRS, 2022, p. 21; Financial Times, 2022; Hunder & Landay, 2022; O'Brian, 2022;). In November, the Ukrainian military operations pushed Russia to retreat across the Dnipro River and beyond (Beaumont & Sauer, 2022; Kirby, 2023). The winter forced the offensive to slow down, with a front line of 1,500 kilometres (January 2023). Large cities in Ukraine faced intense missile bombardments, which intensified as a reaction to the attack on the Kerch bridge in September 2022, and later over Christmas and New Year (Adams, 2022; Kuznetsov, 2022; CNBC, 2023). The Ukrainian counteroffensive continued in spite of constant Russian attacks on the civilian population and critical infrastructure. By the start of 2023, the Commander-in-Chief of the Armed Forces of Ukraine, Zaluzhnyi, announced that Ukraine had liberated 40% of the territory seized by Russia since the full-scale invasion in February 2022. Additionally, the Ukrainian army had liberated 28% of all occupied territories from 2014 onwards (Kirby, 2023; Knight, 2023). Since the Russian invasion, the Ukrainian army has proved resilient, adaptable, and willing to change conditions to exploit adversary military forces (CRS, 2022, p. 11).

The Concept of Memetic War – The Theoretical Foundation

The war against Ukraine quickly became a multidimensional conflict, with conventional military operations, cyberattacks, and information warfare all playing a role. Russia used a range of tactics, including propaganda and disinformation campaigns, to try to shape the narrative of the war and justify its actions to the international community. Ukraine's armed forces, governmental actors, and civic society put up a determined resistance both off- and online. The war in Ukraine has significantly impacted the region and, indeed, the world. It has caused the deaths of thousands of people, created a humanitarian crisis, and sparked a new wave of tensions between Russia and the West. It has also highlighted the growing importance of memetic warfare and the need for countries to develop effective strategies to counter this new form of conflict.

For years, Ukraine has been a testing ground for cyberwarfare, information warfare, and other online attacks. Since 2014, Russia and Russian proxies have targeted Ukraine online, taking advantage of the war in the eastern part of the country to test their cyber offensive weapons (see Chapters 2 and 3) (Cerulus, 2019; Munk, 2022, pp. 126, 131). The 2022 Russian invasion of Ukraine has underscored the importance of cyberwar and information warfare in modern military strategies, which must be considered alongside conventional warfare. Despite the Russian armed forces' advancements in cyber capabilities, Ukraine has quickly developed an innovative and dynamic cyber defence system that involves public and private actors.

Since 2016, the Ukrainian tech industry has boomed, with several young graduates entering the field every year. Numerous multinational companies have established hubs in Ukraine, attracted by the country's reputation as a low-cost offshoring destination (Bruner, 2022; Chakravorti, 2022; Davies, 2022; Sweney, 2022). In spite of the Russian invasion, Ukrainian tech businesses have proved resilient and continue to thrive. Many companies and workers have been actively involved in resistance efforts, including volunteering, fighting in the IT army, and providing financial support to the Ukrainian armed forces. Meanwhile, the Ukrainian IT community has continued contributing to the country's economy by providing innovative services to foreign clients (see Chapter 3) (Prots, 2022; Segal, 2022).

The Ukrainian population has also made significant strides in enhancing its digital literacy and proficiency in computer technologies, networks, and platforms since 2019 (Udovyk et al., 2020; EU NeighboursEast, 2021; UNDP, 2022). During the war, the digitalisation of critical services and communications has proved to be a lifeline for the Ukrainian population and businesses, with digital platforms adapted to provide learning, public services, and other valuable spaces to support the war effort (Ionan, 2022; Interoperable Europe, 2022; Bandura & Staguhn, 2023).

Use of Memes

Originally, memetic war is associated with forcefully circulating memes to reach a political end, i.e. obstructing political success and change (Stiegermark, 2020, p. 110). Regarding online offensive trolls, the provocations are intentional, aggressive, and based on attacking opponents. These actions entail a much broader repertoire: posting untrue opinions, being intently categorical, and making abruptly off-topic comments (see Chapter 3) (Lavorgna, 2020, p. 95; Lindgren, 2022, p. 144; Yar & Steinmetz, 2019, p. 161). Previous wars and conflicts have not understood memetic warfare and how to capitalise on social media and visual platforms' structure and popularity. Social media, virtual spaces, and online communication have not been considered a vital battlefield similar to physical territory or cyberspace, where hacking, data compromises, and spying are widely used. These means and methods have primarily been deployed to damage, destroy or distort data or networks of the adversary state (Giesea, 2015, p. 71; Munk, 2022, p. 49). The Kremlin's use of offensive online trolls, communications and memes for propaganda and dis-and-misinformation during the annexation of Crimea, the war in the Donetsk and Luhansk regions, and beyond have called for actions to debunk and counter these messages as the communication stream intensified after the full-scale invasion (see Chapters 2 and 3) (Gramer, 2017).

The collaboration in memetic warfare is an informal, unstructured, and practical process, with some routine actions such as circulating defensive visual and non-visual communications online and rebuking Russian actors and supporters' online communications. The online users engaged in memetic war have developed accessible and useful response processes, where the collaboration is based on the mutual engagement of participants working together to solve problems (Demchak, 2011, p. 53). Memes are not only about the online users' personal experiences or the story they tell; they are also helpful in bringing people together across time, space, and social media networks. The online environment and memes are keystones in building networks and connecting units of social information (see Chapters 3 and 5) (Sujon, 2021, pp. 17–18).

The Actors

Social identity theory provides a useful framework for understanding how various governmental, military, and civic groups interact online to promote a specific narrative and counter propaganda and dis-and-misinformation (Hogg, 2016, p. 3). The online actors involved in the Ukrainian memetic war have demonstrated their ability to counter Russian propaganda and disinformation through their innovative skills. Defensive meme actors have developed successful countermeasures by debunking false information in posts and comments, spreading news, fact checking, informing, and engaging online users (see Chapter 3). These memetic communities have increased resistance and fostered a sense of togetherness based on a shared experience of inclusion (Ukrainians/Ukrainian supporters) and exclusion (Russians/Russian supporters) (see Chapter 4) (Ross & Rivers, 2017, p. 3; Mortensen & Neumayer, 2021, p. 2368).

In-group members actively promote distinctiveness as they are self-defined and evaluated internally. In doing so, they collectively assess members' status, prestige, and social values in accordance with the group (Hogg, 2016, p. 9). In the case of the Ukrainian memetic war, the groups and individuals involved include internal actors and online users from global networks who support Ukraine. President Zelenskyy's online communication skills have proved effective in reaching various audiences, including the Ukrainian population, international actors, parliaments, governments, organisations, and individuals. His constant stream of video recordings and live broadcasts has created a playbook for communication warfare in the future (see Chapter 5) (Dyczok & Chung, 2022, p. 147; Munk & Ahmad, 2022).

Theoretical Considerations

There is a growing trend of ad hoc collaborations between governmental, military, and civic groups and individuals on social media networks. These collaborations aim to counter Russia's offensive information flow by communicating with in-group members and global actors (see Chapters 3 and 4). However, there is limited research available to provide insights into the concept of memetic war, its parameters, and the actors involved. To fully understand this concept, it is important to unpack the societal, historical, and political approaches that may be at play (see Chapter 2). This requires conducting further research and analysis of the phenomenon. Furthermore, military strategies are typically developed within the civil-military sphere and are linked to the relationship between political and military leadership (see Chapters 2 and 3) (Andersen & McDonald Snookermany, 2020, p. 133). As such, any efforts to address a memetic war will likely involve civilian and military authorities. Overall, gaining a deeper understanding of memetic war and its potential implications for society, politics, and national security is crucial. This requires continued research and collaboration across various sectors and disciplines.

The war against Ukraine has indeed highlighted the crucial role that communication and technology can play in military and political strategies. Online platforms have become a crucial battleground for shaping public opinion, disseminating information, and mobilising support for various causes. Governments and military actors recognise the importance of these platforms and have started directing their online actions and communications to achieve their objectives. However, they are not the only actors involved in this conflict. There is also a voluntary online army made up of individuals driven by a sense of justice and willing to use their technological skills to fight for their cause.

This voluntary army is autonomous and fragmented, with varying levels of engagement, skills, and incitement to be involved. Nevertheless, their impact should not be underestimated. They have become a formidable force in shaping the narrative of the conflict, creating new communication pathways and processes, and forming constellations of like-minded individuals (see Chapter 4) (Drummond, 2022; KnowYourMeme, 2022; North Atlantic Fella Organization, 2022; Ukrainer, 2022; Ukrainian Memes Forces, 2022). In this context, the new weapon is people's technological skills and engagement in the conflict. It is a reflection of the growing power of civil society and the emergence of new forms of civic engagement in the digital age. The Ukrainian conflict has shown that non-governmental actors can play a significant role in shaping the defence in a conflict. Governments and military actors should work to engage with civil society actors and empower them to contribute to conflict resolution efforts. Top of Form

Structural and Network Approaches

Memetic warfare is closely linked to structuration theory, which posits that memes are created, shared, imitated, remixed, iterated, and distributed through dynamic interactions between agency and structure. The included structure refers to the interactions between agency and key events, where memetic warfare links the situation and the motive for engaging in the memetic warfare (Giddens, 1979; McPhee et al., 2014, p. 75). The online structure creates a pattern of interactions between numerous micro-actors (individuals), meso-actors (groups), and macro-actors (governmental authorities and institutions) depending on the communication and actions (McPhee et al., 2014, p. 75). The structuration theory encompasses two important aspects that aid in understanding the groupings and actions within these groups and networks. The concept of structuration includes double hermeneutics that operates in intentional interactions among online users, primarily stemming from shared understanding and perspectives on the war/support for Ukraine. Using particular rhetoric amplifies the understanding and interactions between groups and individuals. Different online users can understand one another, in spite of coming from diverse backgrounds, cultures, and countries, using different codes, symbols, and speeches. These communication tools act as the glue between groups and individuals. Online users interpret one another's behaviour, enabling effective communication and understanding (Leydesdorff, 2010, p. 2139).

The network of actors is complex and non-linear, where the communications and groups constantly change depending on the social interactions and the events during the war. The complexity theory provides a valuable framework for understanding the dynamic interactions and behaviours of the actors involved in memetic warfare and highlights the importance of the flexibility and adaptability of the networks and communities involved in these activities (Munk, 2022, p. 84; Sammut-Bonnici, 2015, p. 1). Complexity theory is inbuilt into the framework as it uses the study of complex systems and behaviours related to strategies and movements by exploring behaviours that is uncertain and non-linearly (Ganco, 2015, p. 1; Munk, 2022, p. 84).

These users exploit their substantial network capacities to reinforce particular messages and range beyond the ability of traditional communication routes. This social media structure of interconnected nodes enables them to reach a large group of people and continue a constant high-level, low-cost communication agenda based on shared values and goals (see Chapter 5). Short messages, visual images, and catchy one-liners attract attention and trigger an emotional response, ultimately creating a sense of community and shared identity among groups and individuals. Visual memes make these communications more easily digestible and memorable, leading to higher rates of retention and repetition. The complex and non-linear nature of these interactions and communications makes it difficult for traditional institutions and governments to control or manipulate the message and the outcome of the memetic war. Instead, the power lies in the hands of the online users, who can create and shape the narrative based on their collective efforts and shared goals.

Overall, the application of social identity theory, structuration theory, and complexity theory provides a comprehensive framework for understanding the dynamics and impact of memetic warfare in the context of the Ukrainian conflict. These theories enable a framework to analyse the interactions between online actors, the role of shared identity and communication strategies, and the importance of network structures and complexity in the success of memetic communications.

Aim and Scope of the Book

The central perspective of this book is memetic war, included in the cyberwar and information warfare framework. A thematic analysis of current events is used to develop an understanding of the parameters included in utilising this concept in Ukraine as part of civic resistance. However, little has been done to develop and apply social media and memes directly to defensive war strategies, where the concepts have primarily been understood in the context of far-right extremism and terrorism.

Although memetic warfare might not have been formalised as an independent strategy during Russia's war against Ukraine, the informal and loosely formed structure is evolving. From the Ukrainian side, initiatives have emerged using memes as a defence tactic to uphold the information flow about the war and to debunk propaganda and dis-and-misinformation. The war has enhanced the visibility of governmental and civic actions and communications using the advantages of social media architecture, networks, and pathways. This book aims to shed insight into the means and methods used in memetic warfare to understand the underlying reasons for engagement, and resistance. The book also considers areas that have strengthened this warfare model and propose areas necessary to include in a coherent memetic defence strategy.

The Research Method

The research conducted in social science adheres to all ethical rules for scientific research in criminology. The research has been conducted objectively and critically regarding data collection, sampling, and analysis. The method for data collection is a combination of primary and secondary research. The research is quantitative, as the text, images, and context are essential, not the numerical data. The online sources were identified and selected using relevant keywords related to events, actors, institutions, and groups. The time for data collection is 12 months, from 24 February 2022, to 23 February 2023.

Secondary Research

The secondary research is based on desk-based grounded methods, using academic books, book chapters, peer-reviewed articles, reports, legislation, policy papers, online websites, social media accounts, and news outlets. As the war unfolds while the research has progressed, there are limited academic sources; the study relies on news reporting and the few academic articles and books published in 2022 about the war. Therefore, the research includes excessive use of online news outlets and websites for thematic data collection, where other sources have validated each argument, and only trustworthy sources have been used. Information has been gathered predominantly online and sampled in themes useful for analysing memetic warfare in conjunction with the Twitter memes collected during primary research. The lens of the research and the arguments are based on Ukraine, and Ukrainian supporters' use of memes as a defensive tool. Therefore, there is no engagement with the meme culture in Russia and among Russian supporters.

The author has no connection to Ukraine or the region and does not have any Ukrainian language skills. However, there are no significant language barriers concerning this research area. The Ukrainian parliament, Verkhovna Rada of Ukraine, the armed forces, the government, and governmental actors publish various information online in Ukrainian and English; this includes official documents, legislation, and strategies. The memes, videos, and communications by Ukrainian authorities, including the Ministry of Defence, Commander-in-Chief of the Armed Forces of Ukraine, the Ministry of Foreign Affairs, the Ukrainian parliament, and the Office of the President of Ukraine, are published in both Ukrainian and English. Moreover, individuals online create memes and communicate predominantly in English, as the online environment is based on internal and external groups and individuals. The Ukrainian Memes Force and North Atlantic Fella Organization (NAFO) predominantly post memes in English – very few will have both English and Ukrainian texts. News outlets such as *The Kyiv Independent* and *Euromaidan Press* are published in English.

Primary Research

The primary data for this research has been collected from social media sites using Twitter. Individuals, groups, and institutions worldwide use Twitter, and the platform's architecture makes it helpful for reaching out to many people. The volume of users, tweets, and hashtags makes Twitter a valuable social media research platform for quantitative data analysis. Through digital ethnography, it is possible to gain insight into social norms, behaviours, patterns, and phenomena that are appropriate to understand the social and political context of memetic communication (Kaur-Gill & Dutta, 2017; Lindgren, 2022, p. 241). Digital ethnography is used to observe the relationship between communications, online groups, and individuals and how these actors can progress a defensive structure internally and externally (Kozinets, 2015, pp. 36–37; Kaur-Gill & Dutta, 2017). Memes are valuable sources for analysing data about behaviour on social media. Memes are:

- Cultural expressions transmitted through the use of visual and non-visual material.
- Civic resistance through a human-initiated process that informs and reflects society and critical events.
- Distribution and redistribution of communications, visual and non-visual, and behaviour on social media.

As with any research using social media data, ethical implications have been considered to ensure that data is collected and used responsibly and transparently. The author has taken measures to ensure the security and privacy of social media users while researching Twitter. Anonymisation of tweets is a common practice in social media research to protect the identities of individuals and ensure confidentiality. It is also important to note that the use of memes as a source of data can be valuable in understanding cultural expressions and social phenomena. The author is also aware of the potential limitations and biases of using them as data sources.

The tweets used have been anonymised to ensure the security of both the memesters and social media users. Although the tweets have been published in the public forum, the authors of the different tweets and posts are protected from being involuntarily exposed in a book for security reasons (Williams et al., 2017, p. 1164; Lindgren, 2022, p. 252). The tweets from ordinary users are not directly discussed or referenced in the text, except for a general thematic analysis of the memes and communications. However, governmental

Table 1.3 Outline of Meme Areas Identified During Data Collection

Research areas	The memes observed and used in this research
One	Everyday communication
Two	Valuable speeches, cultural symbols, and codes
Three	Creation and distribution of memes
Four	Government and non-governmental agencies and actors (internally and externally)
Five	Infuse agency, humour and, creativity into a defence strategy
Six	Influence and being influenced by memes

accounts with the grey badge verifying that they are official public accounts are referenced in the text. Moreover, for most of the memes and online posts in comment fields, it is irrelevant to know who posted them. Instead, the content and how these comments link into overall themes is relevant.

Outline of the Book

This book is about Ukraine's expansion as a memetic warfare superpower within a relatively short period of time. The first chapter conceptualises the current Russian full-scale war against Ukraine, launched on 24 February, 2022. Moreover, this chapter outlines the concept of memetic war and the theoretical and methodological foundation for the research. Chapter 2 provides insight into key historical events and the Russian justification for launching a full-scale invasion, whereas Chapter 3 creates a fundament for understanding the key parameters of war, cyberwar/information warfare, and memetic war. This chapter includes regulation, strategies, propaganda and dis-and-misinformation, and a definition of memetic war.

This leads to two case studies in Chapters 4 and 5. Chapter 4 investigates the agents and actors involved in memetic warfare. Civic engagement, resistance, and togetherness are essential to understanding the interactions and the incitement for engaging in memetic war. Chapter 5 focuses on memes and their content. This chapter relates to behaviours and communications, language, and symbols. The chapter also links the discussions to key features identified in the investigations of memes on Twitter. Finally, Chapter 6 concludes the case study of the Ukrainian defensive memetic war with strategic considerations for future use of the concept. This provides insight into the core elements of memetic warfare, what it aims to cover, and how it would work in an informal and decentralised defence strategy.

References

Adams, P., 2022. *Crimean bridge: Excitement and fear in Ukraine after bridge blast.* [Online] Available at: https://www.bbc.co.uk/news/world-europe-63183409 [Accessed 06 01 2023].

Andersen, M. & McDonald Snookermany, A., 2020. The Making of Military Strategy. The Gravity of an Unequal Dialogue. In *Military Strategy in the 21st Century. The Challenge for NATO.* London: Hurst, pp. 131–151.

Bandura, R. & Staguhn, J., 2023. *Digital will drive Ukraine's modernization.* [Online] Available at: https://www.csis.org/analysis/digital-will-drive-ukraines-modernization [Accessed 25 03 2023].

Barrington, L., 2022. *Putin's key mistake? Not understanding Ukraine's blossoming national identity – even in the Russian-friendly southeast.* [Online] Available at: https://theconversation.com/putins-key-mistake-not-understanding-ukraines-blossoming-national-identity-even-in-the-russian-friendly-southeast-183576 [Accessed 18 09 2022].

BBC News, 2022a. *Capitol riots timeline: What happened on 6 January 2021?* [Online] Available at: https://www.bbc.co.uk/news/world-us-canada-56004916 [Accessed 21 03 2023].

BBC News, 2022b. *Ukraine conflict world reaction: Sanctions, refugees and fears of war.* [Online] Available at: https://www.bbc.co.uk/news/world-europe-60507016 [Accessed 21 03 2022].

BBC News, 2022c. *Ukraine war: Who is winning?* [Online] Available at: https://www.bbc.co.uk/ news/explainers-62902029 [Accessed 17 09 2022].

Beaumont, P. & Sauer, P., 2022. *Russian troops ordered to retreat from Kherson in face of Ukrainian advance.* [Online] Available at: https://www.theguardian.com/world/2022/nov/09/russians-destroy-dnieper-bridges-to-slow-ukraine-advance-on-kherson [Accessed 06 01 2023].

Bruner, R., 2022. *"Now I'm working twice as hard." How one Ukrainian is working remotely through war.* [Online] Available at: https://time.com/6160481/ukraine-remote-workers/ [Accessed 03 04 2022].

CBS News, 2022. *Russia's Putin announces annexation of 4 Ukraine regions, despite global outcry.* [Online] Available at: https://www.cbsnews.com/news/ukraine-russia-putin-annexes-four-regions-says-will-defend-territory-by-all-means/ [Accessed 06 01 2023].

Cerulus, L., 2019. *How Ukraine became a test bed for cyberweaponry.* [Online] Available at: https://www.politico.eu/article/ukraine-cyber-war-frontline-russia-malware-attacks/ [Accessed 26 03 2023].

Chakravorti, B., 2022. *Ukraine invasion disrupts a vital tech talent pool.* [Online] Available at: https://www.bloomberg.com/opinion/articles/2022-02-28/ukraine-invasion-will-disrupt-another-part-of-the-tech-sector-it-workers [Accessed 03 04 2022].

CNBC, 2023. *Defiant Ukrainians cheer New Year as drones blasted from skies.* [Online] Available at: https://www.cnbc.com/2023/01/01/defiant-ukrainians-cheer-new-year-as-drones-blasted-from-skies.html [Accessed 03 01 2022].

CRS, 2022. *Russia's war in Ukraine: Military and intelligence aspects,* Washington, DC: Congressional Research Service.

David, D., 2022. *What impact has Brexit had on the UK economy?* [Online] Available at: https://www.bbc.co.uk/news/business-64450882 [Accessed 21 03 2023].

Davidson Sorkin, A., 2021. *What Angela Merkel left behind.* [Online] Available at: https://www.newyorker.com/news/daily-comment/angela-merkel-leaves-politics-on-her-own-terms [Accessed 21 03 2023].

Davies, P., 2022. *Ukraine's tech sector is a "pillar of resistance". Here's how it's responding to Russia's invasion.* [Online] Available at: https://www.euronews.com/next/2022/02/25/ukraine-s-tech-sector-is-a-pillar-of-resistance-here-s-how-it-s-responding-to-russia-s-inv [Accessed 30 03 2022].

Demchak, C.C., 2011. *Wars of Disruption and Resilience.* Athens: University of Georgia.

Donovan, J., Dreyfuss, E., & Friedberg, B., 2022. *Meme Wars.* New York: Bloomsbury Publishing.

Drummond, M., 2022. *Ukraine's internet army of "fellas" are using dog memes to fight Russian propaganda – and they've raised $1m for the army too.* [Online] Available at: https://news.sky.com/story/ukraines-internet-army-of-fellas-are-using-dog-memes-to-fight-russian-propaganda-and-theyve-raised-1m-for-the-army-too-12729625 [Accessed 26 12 2022].

Dyczok, M. & Chung, Y., 2022. Zelens'kyi Uses his Communication Skills as a Weapon of War. *Canadian Slavoniv Papers*, 64(2–3), pp. 146–161.

Economist, The, 2022. *An interview with General Valery Zaluzhny, head of Ukraine's armed forces*. [Online] Available at: https://www.economist.com/zaluzhny-transcript [Accessed 06 01 2023].

Epstein, J. & David, C.R., 2022. *Putin thought Russia's military could capture Kyiv in 2 days, but it still hasn't in 20*. [Online] Available at: https://www.businessinsider.com/vladimir-putin-russian-forces-could-take-kyiv-ukraine-two-days-2022-3?r=US&IR=T [Accessed 18 09 2022].

EU NeighboursEast, 2021. *Ready for the digital decade? Improving skills to meet the technological challenge in Ukraine*. [Online] Available at: https://euneighbourseast.eu/news/stories/ready-for-the-digital-decade-improving-skills-to-meet-the-technological-challenge-in-ukraine/ [Accessed 25 03 2023].

Financial Times, 2022. *Russia's invasion of Ukraine in maps — latest updates*. [Online] Available at: https://www.ft.com/content/4351d5b0-0888-4b47-9368-6bc4dfbccbf5 [Accessed 17 09 2022].

Fisher, M., 2022. *Putin's case for war, annotated*. [Online] Available at: https://www.nytimes.com/2022/02/24/world/europe/putin-ukraine-speech.html [Accessed 20 03 2022].

Ganco, M., 2015. Complexity Theory. In *Wiley Encyclopedia of Management*. London: Wiley.

Giddens, A., 1979. *Central Problems in Social Theory. Action, Structure and Contradiction in Social Analysis*. London: Palgrave Macmillan.

Giesea, J., 2015. It's Time to Embrace Memetic Warfare. *Defence Strategic Communications*, 1(1), pp. 67–75.

Global Conflict Tracker, 2022. *Conflict in Ukraine*. [Online] Available at: https://www.cfr.org/global-conflict-tracker/conflict/conflict-ukraine [Accessed 03 01 2022].

Gramer, R., 2017. *Can NATO weaponise memes?* [Online] Available at: https://foreignpolicy.com/2017/04/13/nato-cyber-information-warfare-battle-of-ideas-memes-internet-culture/ [Accessed 26 12 2022].

Harari, Y.N., 2022. *Why Vladimir Putin has already lost this war*. [Online] Available at: https://www.theguardian.com/commentisfree/2022/feb/28/vladimir-putin-war-russia-ukraine [Accessed 18 09 2022].

Harding, L., 2022. *Demoralised Russian soldiers tell of anger at being "duped" into war*. [Online] Available at: https://www.theguardian.com/world/2022/mar/04/russian-soldiers-ukraine-anger-duped-into-war [Accessed 19 09 2022].

Harris S., DeYoung, K., Khurshudyan, I., Parker, A., & Sly, L. 2022. *Road to war: U.S. struggled to convince allies, and Zelensky, of risk of invasion*. [Online] Available at: https://www.washingtonpost.com/national-security/interactive/2022/ukraine-road-to-war/?itid=lb_war-in-ukraine-what-you-need-to-know_3 [Accessed 23 12 2022].

Herszenhorn, D.M. & McLeary, P., 2022. *Ukraine's "iron general" is a hero, but he's no star*. [Online] Available at: https://www.politico.com/news/2022/04/08/ukraines-iron-general-zaluzhnyy-0002390 [Accessed 05 01 2022].

Hogg, A., 2016. Social Identity Theory. In *Understanding Peace and Conflict Through Social Identity Theory*. Cham: Springer Nature, pp. 3–18.

Hunder, M. & Landay, J., 2022. *Russia launches biggest air strikes since start of Ukraine war*. [Online] Available at: https://www.reuters.com/world/europe/russias-ria-state-agency-reports-fuel-tank-fire-kerch-bridge-crimea-2022-10-08/ [Accessed 06 01 2023].

Interoperable Europe, 2022. *Digital Public Administration Factsheet 2022*. Brussels: Interoperable Europe.

Ionan, V., 2022. *Digital transformation in Ukraine: Before, during, and after the war*. [Online] Available at: https://www.sir.advancedleadership.harvard.edu/ articles/ digital-transformation-in-ukraine-before-during-after-war [Accessed 25 03 2023].

Kaur-Gill, S. & Dutta, M., 2017. Digital Ethnography. In *The International Encyclopedia of Communication Research Methods*. London: Wiley, pp. 1–11.

Kirby, P., 2022. *What does Putin want and will Russia end its war?* [Online] Available at: https://www.bbc.co.uk/news/world-europe-56720589 [Accessed 17 03 2022].

Kirby, P., 2023. *Makiivka: Russia points fingers after deadliest Ukraine attack*. [Online] Available at: https://www.bbc.co.uk/news/world-europe-64155859 [Accessed 06 01 2023].

Knight, M., 2023. *Ukrainian military chief says 40% of territories occupied during Russian invasion now liberated*. [Online] Available at: https://edition.cnn.com/europe/ live-news/russia-ukraine-war-news-1-3-23/h_43aea12245bf3bf98fb78f3209631199 [Accessed 06 01 2022].

KnowYourMeme, 2022. *NAFO/North Atlantic Fella Organization*. [Online] Available at: https://knowyourmeme.com/memes/cultures/nafo-north-atlantic-fella-organization [Accessed 17 09 2022].

Kozinets, R., 2015. *Netography: Redefined*. 2nd ed. London: Sage.

Kuznetsov, S., 2022. *Christmas Eve missile strike kills at least 8 people in Ukraine city of Kherson*. [Online] Available at: https://www.politico.eu/article/christmas-eve-missile-strike-kills-7-ukraine-kherson/ [Accessed 06 01 2023].

Lavorgna, A., 2020. *Cybercrimes. Critical Issues in a Global Context*. London: Macmillan.

Leicester, J., 2022. *To Europe's relief, France's Macron wins but far-right gains*. [Online] Available at: https://apnews.com/article/2022-french-election-voting-results-f5b549e3b99930ee05bed17a9d3870b6 [Accessed 21 03 2023].

Leydesdorff, L., 2010. The Communication of Meaning and the Structuration of Expectations: Giddens' "structuration theory" and Luhmann's "self-organisation." *Journal of the American Society for Information Science & Technology*, 61(10), pp. 2138–2150.

Lindgren, S., 2022. *Digital Media and Society*. 2nd ed. London: Sage.

Liptak, K., 2022. *Biden says Russia is beginning an "invasion of Ukraine" as he unveils sanctions on Moscow*. [Online] Available at: https://edition.cnn.com/2022/02/22/ politics/biden-invasion-sanctions/index.html [Accessed 27 02 2022].

Lister, T., John, T., & Murphy, P.P., 2022. *Here's what we know about how Russia's invasion of Ukraine unfolded*. [Online] Available at: https://edition.cnn.com/2022/ 02/24/europe/ukraine-russia-attack-timeline-intl/index.html [Accessed 17 03 2022].

Lutska, V., 2022. *Ukrainian meme forces: What makes us laugh in the times of Russia's invasion*. [Online] Available at: https://war.ukraine.ua/articles/ukrainian-meme-forces-what-makes-us-laugh-in-the-times-of-russia-s-invasion/ [Accessed 23 11 2022].

Marson, J., 2022. *Putin thought Ukraine would fall quickly. An airport battle proved him wrong*. [Online] Available at: https://www.wsj.com/articles/putin-thought-ukraine-would-fall-quickly-an-airport-battle-proved-him-wrong-11646343121 [Accessed 19 09 2022].

McPhee, R., Poole, M., & Iverson, J., 2014. Structuration Theory. In *The SAGE Handbook of Organisational Communication: Advances in theory, research, and methods*. Thousand Oaks: Sage, pp. 75–110.

Mortensen, M. & Neumayer, C., 2021. The Playful Politics of Memes. *Information, Communication and Society*, 24(16), pp. 2367–2377.

Munk, T., 2022. *The Rise of Politically Motivated Cyber Attacks.* London: Routledge.

Munk, T. & Ahmad, J., 2022. "I Need Ammunition, Not a Ride": The Ukrainian cyber war. *Comunicação e Sociedade*, 42, pp. 221–241.

NATO, 2022. *Relations with Russia.* [Online] Available at: https://www.nato.int/cps/en/natolive/topics_50090.htm [Accessed 21 03 2022].

North Atlantic Fella Organization, 2022. *@Official_NAFO.* [Online] Available at: https://twitter.com/Official_NAFO/status/1577269391997050880?ref_src=twsrc%5Etfw%7Ctwcamp%5Etweetembed%7Ctwterm%5E1577274895095906305%7Ctwgr%5E60338d1736a091a06999d94fc562e0c9cc56eff2%7Ctwcon%5Es2_&ref_url=https%3A%2F%2Fnews.sky.com%2Fstory%2Fukraines-int [Accessed 27 12 2022].

O'Brian, P.P., 2022. *Ukraine pulled off a masterstroke.* [Online] Available at: https://www.theatlantic.com/ideas/archive/2022/09/ukraine-russia-putin-kharkiv-kupyansk/671407/ [Accessed 18 09 2022].

O'Connor, M., 2022. *Russia attack on Ukraine catastrophe for Europe, say Boris Johnson.* [Online] Available at: https://www.bbc.co.uk/news/uk-60504204 [Accessed 27 02 2022].

Prots, U., 2022. *Ukrainian tech industry overview in 2022.* [Online] Available at: https://techreviewer.co/blog/ukrainian-tech-industry-overview-in-2022 [Accessed 25 03 2023].

Ross, A.S. & Rivers, D.J., 2017. Digital Cultures of Political Participation: Internet memes and thediscursive delegitimisation of the 2016 US presidential candidates. *Discourse, Context & Media*, 6, pp. 1–11 (https://www.sciencedirect.com/science/article/abs/pii/S2211695816301684? via%3Dihub).

Sabbagh, D., 2022. *Ukraine continues Kharkiv offensive despite apparent Russian retaliation.* [Online] Available at: https://www.theguardian.com/world/2022/sep/12/ukraine-continues-kharkiv-offensive-despite-apparent-russian-retaliation [Accessed 26 03 2023].

Sammut-Bonnici, T., 2015. Complexity Theory. In *Wiley Encyclopedia of Management.* London: Wiley, pp. 1–2.

Sauer, P. & Harding, L., 2022. *Putin annexes four regions of Ukraine in major escalation of Russia's war.* [Online] Available at: https://www.theguardian.com/world/2022/sep/30/putin-russia-war-annexes-ukraine-regions [Accessed 06 01 2023].

Segal, D., 2022. *Inside Ukraine's thriving tech sector.* [Online] Available at: https://www.nytimes.com/2022/07/22/business/ukraine-tech-companies-putin.html [Accessed 25 03 2023].

Smith, D., 2022. *Biden's Russia sanctions: Why holding back could be part of his strategy.* [Online] Available at: https://www.theguardian.com/us-news/2022/feb/22/joe-biden-russia-first-sanctions-ukraine-analysis [Accessed 21 03 2022].

Stiegermark, A., 2020. *Memetic Warfare.* Helsingborg: Logik.

Sujon, Z., 2021. *The Social Media Age.* 1st ed. London: Sage.

Sweney, M., 2022. *Keyboard warriors: Ukraine's IT army switches to war footing.* [Online] Available at: https://www.theguardian.com/world/2022/mar/12/keyboard-warriors-ukraines-it-army-switches-to-war-footing [Accessed 27 03 2022].

Udovyk, O., Moskalenko, O., & Kylymnyk, I., 2020. *Bridging the digital divide in Ukraine: A human-centric approach.* [Online] Available at: https://www.undp.org/

ukraine/blog/bridging-digital-divide-ukraine-human-centric-approach [Accessed 25 03 2023].

Ukrainer, 2022. *Who are the NAFO Fellas? The army of cartoon dogs fighting Russian propaganda.* [Online] Available at: https://ukrainer.net/nafo-fellas/ [Accessed 27 12 2022].

Ukrainian Memes Forces, 2022. *@uamemesforce.* [Online] Available at: https://mobile.twitter.com/uamemesforces [Accessed 01 01 2023].

UNDP, 2022. *Digital education for older persons extended all over Ukraine.* [Online] Available at: https://bearr.org/regional-news/digital-education-for-older-persons-extended-all-over-ukraine/ [Accessed 25 03 2023].

UNIS, 2022. *Crisis in Ukraine.* [Online] Available at: https://unis.unvienna.org/unis/en/events/2022/crisis-in-ukraine.html [Accessed 21 03 2022].

United Nations, 2022. *Secretary-General says Russian Federation's recognition of "independent" Donetsk, Luhansk violate Ukraine's sovereignty, territorial integrity.* [Online] Available at: https://www.un.org/press/en/2022/ sgsm21153.doc.htm [Accessed 27 02 2022].

WEF, 2022. *These charts show Europe's reliance on gas before the war in Ukraine.* [Online] Available at: https://www.weforum.org/agenda/2022/11/europe-gas-shortage-russia/ [Accessed 21 03 2023].

White House, The, 2022. *Remarks by President Biden announcing response to Russian actions in Ukraine.* [Online] Available at: https://www.whitehouse.gov/briefing-room/speeches-remarks/2022/02/22/remarks-by-president-biden-announcing-response-to-russian-actions-in-ukraine/[Accessed 21 03 2022].

Williams, M., Burnap, P., & Sloan, L., 2017. Towards an Ethical Framework for Publishing Twitter Data in Social Research: Taking into account users' views, online context and algorithmic estimations. *Sociology*, 51(6), pp. 1149–1168.

Yar, M. & Steinmetz, K.F., 2019. *Cybercrime and Society*. 3rd ed. London: Sage.

Ziady, H., 2022. *Brexit has cracked Britain's economic foundations.* [Online] Available at: https://edition.cnn.com/2022/12/24/economy/brexit-uk-economy/index.html [Accessed 21 03 2023].

2   Tensions and False Justifications

Escalation of Events

Tine Munk

Revolutions and War

On Christmas Day, 1991, Soviet leader Gorbachev announced on TV that the Soviet Union (the Union of Soviet Socialist Republics (USSR)) no longer existed. Instead, the Union was replaced by 15 separate countries (BBC News, 2016; Myre, 2021). During the decades of Soviet central rule, Ukraine, Georgia, Belarus, Armenia, Azerbaijan, Kazakhstan, Kyrgyzstan, Moldova, Turkmenistan, Tajikistan, and Uzbekistan were seen as one group within the state, and nationhood was institutionalised on all levels of life, including people and the concept of autonomous political units, such as national states. Therefore, states were considered ethnocultural rather than civic entities. After gaining independence, they had to develop their identity free from the suppressive Soviet regime (Kulyk, 2016, p. 591). Following the collapse of the Soviet Union, new power systems emerged in the former republics. The communist-oriented apparatus was dismantled, and political parties participated in elections. However, in spite of the changes, not all states looked towards the West to develop a capitalist democracy. Instead, they adopted different variations of democratic power system (Lane, 2008, p. 525). During the Ukrainian Revolution on the Granite in 1990–1991, students asserted their public right to protest in Kyiv's main square. The protest aimed to remove the head of the Cabinet of Ministers and called for independence (Yekelchyk, 2020, p. 5; Onuch & Hale, 2022, p. 38).

The Orange Revolution

In 2004–2005, the Orange Revolution and the later Euromaidan Revolution (2013–2014) took a central stage on the same Independence Square [Maidan Nezalezhnosti] in Kyiv (Klid, 2007, p. 118; Dickinson, 2022). During the first revolution, the protesters prevented the Kremlin-backed candidate Yanukovych from stealing the Ukrainian presidency in a staged election. The Orange Revolution enabled the Ukrainian people to elect Yanukovych's rival, the more reform-friendly and European-oriented politician, Yushchenko (Lane, 2008, p. 527; D'Anieri, 2019, p. 127; Dickinson, 2020). The revolution

DOI: 10.4324/9781003432630-2

successfully brought people together, calling for democracy, transparency, and change. The protesters wanted the country to move away from cronyism and coercion and transform into a society in which progressive, democratic ideas and a public spirit could develop freely without suppression (Katchanovski, 2008; Onuch & Hale, 2022, p. 78). However, President Yushchenko failed to achieve his aims due to infighting in the government. Instead of bringing in a new team of dedicated reformers, Yushchenko created a broad coalition of established politicians who belonged to the elite and were part of the original problem. They all expected to gain influence, and each had a different vision of the reform process (Reuters, 2007; Katchanovski, 2008; NBCNews, 2008; Onuch & Hale, 2022, p. 87).

The Orange Revolution was a watershed moment in the disintegration from Russia and the Soviet Union's past colonisation of Ukraine, which suppressed its nationalist belief and identity (Lane, 2008, p. 527). Ukrainians called for changes to move away from the manipulations of the old regime and pursue a new democratic pathway and independence (Onuch & Hale, 2022, p. 78). Russia saw this move as a threat to its regional power position, and losing influence over Ukraine would be a geopolitical defeat (Horvath, 2011, p. 6). Russia believed that Western powers, particularly the USA, encouraged civil unrest to diminish Russia's power position (Haaland Matlary & Johnson, 2020, p. 13). After the Orange Revolution, Ukrainian politicians registered a record low approval rate grounded in complex political issues and fears of increased pressure from Russia. Former presidential candidate Yanukovych appeared in the 2010 presidential election as the candidate who could solve the deadlock (Onuch & Hale, 2022, p. 101). Although he was elected this time, his regime was characterised by a multi-pronged approach of concentrating power, similar to the earlier (and very unpopular) President Kuchma. President Yanukovych used his power to control parliament, change laws that restrained him, and amend the constitution to invalidate the amendments adopted in the aftermath of the Orange Revolution. Finally, President Yanukovych consolidated his control over economic resources and increased pressure on Ukrainian businesses (Yekelchyk, 2020, pp. 176–178).

The Euromaidan Revolution

The Euromaidan Revolution, also known as the Revolution of Dignity, began as peaceful protests by the Ukrainian population who were unhappy with Yanukovych's decision to withdraw from the EU Association Agreement (Krushelnycky, 2013; Shveda & Park, 2016, p. 85; Yekelchyk, 2020, p. 161; Dyczok, 2022; Onuch & Hale, 2022, p. 117). The Ukrainian population wanted the deal signed to rebuild their economy through free trade accords, but they also saw this as a starting point for a new prosperous era. With this agreement, Ukraine could defend its fragile democracy and improve areas such as human rights, freedom of the press, etc. (Krushelnycky, 2013).

The Ukrainian people called for a new government, free of corruption and oligarchs, and demanded that their voices be heard. The Ukrainian government's crackdown on protesters during the Euromaidan Revolution of 2013–2014 led to widespread outrage and condemnation both domestically and internationally. The protesters wanted democratic reforms, ending corruption, and closer ties with Europe. The government's violent response, including the use of live ammunition, resulted in the deaths of over 100 protesters and injury to hundreds more (Krushelnycky, 2013; Traynor, 2014; Gorenburg, 2015; Yekelchyk, 2020, pp. 218–219; Onuch & Hale, 2022, p. 120). This brutal repression only fuelled the protesters' determination to continue their struggle for democracy and ultimately led to the ousting of President Yanukovych (BBC News, 2014b; Booth, 2014). The Ukrainian people saw this as a turning point in their history and a decisive break from their authoritarian past. They hoped that the new government would bring the changes they had been fighting for and build a democratic and prosperous future for Ukraine. The revolution resulted in Yanukovych's removal from power and a new government coming into place. However, it also led to the annexation of Crimea by Russia and the beginning of a long-term war in eastern Ukraine.

The Annexation of Crimea and the War in Donetsk and Luhansk

The aftermath of the Euromaidan Revolution was marked by a period of instability and conflict, as Russian-backed separatists seized control of Crimea and pro-Russian rebels rose up in eastern Ukraine. The conflict escalated into a full-scale war, with thousands of lives lost and significant damage to infrastructure and the economy (Pifer, 2020; Yekelchyk, 2020, p. 222). The events of the Euromaidan Revolution and the ongoing war have shaped Ukrainian society and politics, with a renewed emphasis on national identity, democracy, and sovereignty (Kulyk, 2016).

Annexation

The situation in Crimea was precipitated by the political turmoil in Ukraine following President Yanukovych's ousting, which the Crimean elite viewed as a threat to their interests. The propaganda war uses the narrative "neo-Nazi coup" in Kyiv to claim that the situation with the Euromaidan would significantly threaten Crimea's Russian culture and ethnic nationality group. Russia was willing to use force to seize and annex Crimea and wage war using proxies against Ukraine in the Donetsk and Luhansk regions (Yekelchyk, 2020, p. 117).

On 27 February 2014, Russian proxies took control of crucial institutions, facilities, and checkpoints in Crimea. The takeover was conducted by groups of men without identifying insignia, known as the "little green men". This was followed by a request for Russian intervention by the pro-Russian authorities in the region (Chappell & Memmott, 2014; Pifer, 2020; Yekelchyk, 2020). Russia then organised an unconstitutional and illegal referendum to annex the territory. The referendum lacked independent observers to validate the result. This tactic of using a sham referendum to annex territory has also been used by Russia in the 2022 war (BBC News, 2014a; CoE, 2014; D'Anieri, 2019, p. 228; AP, 2022; Harding, 2022).

The War in Donetsk and Luhansk

The annexation of Crimea in March 2014 marked the beginning of a series of events that led to the emergence of Russian organised separatist movements in Ukraine's Donetsk, Kharkiv, and Luhansk regions. Pro-Russian forces seized control of governmental and administrative buildings in these regions, claiming that they rejected Kyiv's authority and sought to create new republics aligned with Russia (Grytsenko, 2014; Yekelchyk, 2020, p. 128; Kingsley & Sommerlad, 2022). The move to establish the Donetsk People's Republic (DPR) and the Luhansk People's Republic (LPR) was met with little resistance in Donetsk, but the Ukrainian authorities regained control over the northern city of Kharkiv. However, so-called separatists continued to make progress, and on 24 May, they declared the formation of the Federation of New Russia, later renamed the Union of People's Republics, to align with Putin's vision of history and Russia's prominent role (Yekelchyk, 2020, pp. 128–129; Kingsley & Sommerlad, 2022).

Indeed, Russia's support for the separatist movements in Donetsk, Kharkiv, and Luhansk went beyond mere encouragement and included military and political support. Russia provided the separatists with organisation, intelligence, funding, and arms, which enabled them to take over cities and political apparatus in the region (D'Anieri, 2019, p. 233; Hutchinson & Reevell, 2022). One critical element in Russia's strategy was to use the "little green men" – attackers dressed in civilian clothes – who made it challenging for the Ukrainian military to distinguish them from the local population (Fiala, 2020, p. 118). As the Ukrainian armed forces started to gain control over the separatists, Russia escalated the conflict by sending in regular troops and heavy weaponry, such as tanks and artillery (Yekelchyk, 2020; Global Conflict Tracker, 2022; Roth, 2022). The war in the region has since then been marked by a prolonged stalemate, with frequent shelling and battles along the frontlines between Russian-backed separatists and Ukrainian forces. The emergence of separatist movements in these regions triggered a protracted conflict. Despite various ceasefire agreements, the situation remained tense, and the ongoing war threatened regional stability. The regional war has resulted in a significant loss of life, with over 14,000 people killed between 2014 and 2022 (Crisis Group, 2022; Hutchinson & Reevell, 2022).

Violations of Agreements

The seizure of Crimea from Ukraine violated several agreements, including the UN Charter, the 1975 Helsinki Final Act, the 1994 Budapest Memorandum of Security Assurances for Ukraine, and the 1997 Treaty on Friendship, Coopera-tion and Partnership between Ukraine and Russia (OSCE, 1975; Sorokowski, 1996; Pifer, 2020; Borda, 2022; UN, 2022). Russia's involvement in the war in eastern Ukraine has been difficult to hide, in spite of its efforts to deny any involvement or downplay its role. Satellite photos, observers from NATO and the OSCE, journalists, think tanks, and investigative units have all captured evi-dence of Russian military personnel, equipment, and weaponry in the region. The Ukrainian military and intelligence services have also documented the pres-ence of Russian combatants. Russia's denials have been further undermined by statements from high-ranking officials, such as Foreign Minister Lavrov's com-ments in 2017, in which he acknowledged Russia's decision to join the fight in Donbas and Syria (Kuzio & D'Anieri, 2018, p. 7). These revelations have fuelled criticism from the international community and increased pressure on Russia to de-escalate the conflict and withdraw its support for the separatist movements.

The Minsk I Agreement created a 12-point ceasefire on 5 September 2014; the deal collapsed shortly after due to heavy fighting in Donetsk Airport. The Minsk II Agreement was agreed to on 15 February 2015 during a heavy battle (BBC News, 2015; D'Anieri, 2019, p. 247; Yekelchyk, 2020, pp. 143–144). The Minsk II Agreement of 2015 aimed at ending the conflict in eastern Ukraine and providing a path towards a peaceful resolution. However, in spite of the agree-ment, Ukraine has not fulfilled its obligations to ensure autonomy for the separa-tist regions and amnesty for rebels, neither has Russia ensured that Ukraine could regain full control of its borders with Russia in these territories (European Par-liament, 2016; Allan, 2020; Kingsley & Sommerlad, 2022). Russia blocked the military and political steps by claiming it was not a part of the conflict and subse-quently not bound by it. The war continued, although the control lines stabilised, and the intensity of the battles was reduced (D'Anieri, 2019, p. 249; Yekelchyk, 2020, p. 162; Reuters, 2022a). Russia's denial of involvement in the conflict has allowed it to continue to utilise its hybrid warfare model, using civilian soldiers and "little green men" to support separatist movements while maintaining plau-sible deniability. This approach has allowed Russia to exert control over events, attack and withdraw without consequences, and maintain a level of uncertainty and unpredictability in the conflict (Fiala, 2020, p. 118; Munk, 2022, p. 124).

Ukraine's Military Doctrine and Reforms

NATO Enlargement

In 1995, NATO moved to enlarge the organisation to the east. The military organisation set the accession criteria, inviting applicant countries to open a dialogue about membership in 1996 (Walker, 2016, p. 139; Yekelchyk, 2020, pp. 88–90). US President Bush's administration pushed NATO member states to offer Membership Action Plans (MAP) to Georgia and Ukraine. The plan was to use the Bucharest Summit in 2008 to allow the two states to start progressing into a full NATO membership.

This plan was vetoed by Germany and France, which feared that this enlargement would be a red line for Russia. A compromise was agreed on to offer future MAPs to the two countries (Erlanger & Myers, 2008; Spetalnick, 2008; Traynor, 2008; Walker, 2016, p. 141; Yekelchyk, 2020, p. 163). The former German Chancellor, Merkel, has been criticised for preventing Ukraine from starting the MAP process. In a short statement after the full-scale war began in 2022, Merkel claimed that she still stands by this decision from 2008, although the membership would have been a barrier to a Russian invasion, as NATO's Article 5 requires a collective response from all member states if one state is attacked (NATO, 1949; France24, 2022; Munk, 2022). It is worth noting that the decision on whether to grant Ukraine MAP was ultimately made by all NATO member states, not just Germany and France. Additionally, while MAP is an important step towards NATO membership, it does not guarantee membership and is only offered to countries that meet specific criteria (NATO, 2008). At the time of these discussions about MAP, Germany believed it was too early for Ukraine to be offered membership due to the political conditions in the country, which it claimed did not meet NATO's requirements (Yekelchyk, 2020, pp. 163–164; Brzozowski, 2022; France24, 2022).

Military Reform

After the 2014 annexation of Crimea, Western countries provided military support to Ukraine, including weapons and training. Among the support provided was technical military assistance to strengthen the Ukrainian military's physical and conceptual components, such as training soldiers in weapons systems and preparing brigades to plan and defend cities (Arnold, 2022). The Ukrainian army suffered a painful defeat at Debaltseve in 2015, where Russian and pro-Russian forces surrounded Ukrainian troops. This defeat and the Minsk II Agreement prompted investments in the Ukrainian Armed Forces (Luhn & Grytsenko, 2015; Noorman, 2020; Yekelchyk, 2020, p. 135).

The Ukrainian military reform programme was crucial for the country's survival in the face of Russian aggression. The programme introduced a new military doctrine prioritising Russia as the main threat, passed by the Verkhovna Rada in 2015 (Verkhovna Rada, 2015). As fighting decreased following the signing of the Minsk II Agreement, Ukraine's Armed Forces had the opportunity to regroup and reconsider their performance. The reform plan aimed to transition the military force into a professional army, with some voluntary detachments moved back from the front and dissolved. Other units were changed into contract regiments as a part of the professional army or came under the Ministry of Internal Affairs (Onuch & Hale, 2022, p. 129;

Yekelchyk, 2020, pp. 135–136). Some of these voluntary units have previously been associated with the far-right movement and controversial behaviour. By being incorporated into the actual military forces, these groups must adhere to the rules of the military command (Onuch & Hale, 2022, p. 129).

The decentralised command structure is also supported by NATO, as it is believed to be more effective in modern warfare, where rapid decision making is essential. This structure allows for faster response times and increased flexibility, which can be crucial in battles. The reforms have also included improvements in logistics and intelligence capabilities, enabling the Ukrainian military to coordinate its operations better and respond to threats (Onuch & Hale, 2022, p. 130). Ukraine's military reform programme has been a significant factor in its ability to withstand Russia's invasion in 2022. The introduction of a new military doctrine, the transition to a professional army, and the decentralisation of the command structure have all played a crucial role in improving the effectiveness of the Ukrainian military. While there is still much work to be done, these reforms have undoubtedly made a difference in Ukraine's ability to defend its territory (Zagoronyuk et al., 2021; Bonenberger, 2022; Herszenhorn & McLeary, 2022; Onuch & Hale, 2022, p. 129;).

Ukrainian soldiers have participated in training programmes with counterparts from NATO and its allies (Yekelchyk, 2020, p. 136; Bonenberger, 2022; NATO, 2022a). The training was seen as vital in improving the performance of Ukrainian soldiers and their ability to counter the Russian invasion. The support from Western nations also provided Ukraine with access to modern military equipment and technology, which has enabled them to counter the advanced weaponry used by the Russian military (Yekelchyk, 2020, p. 136). In 2022, during the full-scale war, personnel and combined arms units have been training on the territory of 17 European counties. This signifies that more than 20,000 soldiers have been trained by partner nations (Commander-in-Chief of the Armed Forces of Ukraine, 2023).

Resilience and Resistance Strategy – Civic Engagement

The Strategic Defence Bulletin (2017) derives from the National Security Strategy of Ukraine and the Military Security Strategy of Ukraine. Under the provisions of the Military Doctrine of Ukraine, a strategic reform programme scheme was launched (see Table 2.1). The Strategic Defence Bulletin aims to transform the Ukrainian Armed Forces into a modern and combat-ready force capable of defending the country's sovereignty and territorial integrity. Implementing the programme has been a top priority for the Ukrainian government, especially after Russia's 2014 annexation of Crimea and the ongoing war in eastern Ukraine. The programme has focused on enhancing the training and understanding of command and control, planning, operations, medical and logistics, and professionalisation of the armed forces (Arnold, 2022). The first part of the reform was implemented during 2016–2020. The second part of

Table 2.1 Strategic Defence Bulletin for Ukraine

Parts	Content
1	Contains effective defence management and a joint defence and military command leadership system. These systems should carry out NATO principles and standards regarding democratic civilian control
2	Develop a professional and motivated personnel defence force. An effective personal management system, education and science, health care and social protection should support trained military reserves
3	Increase the use of modern weapons and military equipment, including special equipment. This is important to ensure that the different forces can achieve tasks set out by the Armed Forces of Ukraine and other components of the defence forces. This is also needed to ensure that Ukraine's armed forces comply with relevant NATO structures
4	Ensure that the other areas can progress. It is essential to develop military infrastructures, logistics and supplies inventory, and a system to provide medical support. These need to be in place, so the Armed Forces of Ukraine and other parts of the defence forces can counter armed aggression against Ukraine.
5	Increase the defence forces' integrated operational (combat and special) capabilities to defend Ukraine against aggressions and hybrid threats

Source: (President of Ukraine, 2021).

the new bulletin sets the plans for 2021–2025 (Atlantic Council, 2014, p. 4; Verkhovna Rada, 2015; GOV.UA, 2020; President of Ukraine, 2021;).

Protection of the integrity of the territory is a duty of every citizen in Ukraine, as established in Article 17 of the Constitution of Ukraine (Verkhovna Rada, 1991; Global Security, 2021). Article 17 covers sovereignty, territorial integrity, and economic and information security. Due to the threat level, it is essential that the Armed Forces of Ukraine can respond and have coordinated capabilities and powers (Verkhovna Rada, 1991). Notable, the Total Defence Concept originated in Sweden during the Cold War and has since been adopted by other countries, including Ukraine (Warsaw Insitute, 2018; Larsson, 2020, p. 45). Total Defence refers to the integration of civilian and military defence efforts to enhance a country's resilience and preparedness in the face of various threats, including armed conflict, natural disasters, and cyberattacks. In the case of Ukraine, the Total Defence Concept includes not only military preparedness but also the participation of civilians in the country's defence efforts (Shelest, 2022). This includes voluntary defence organisations, territorial defence battalions, and civil defence activities, such as emergency response and critical infrastructure protection. The idea is to mobilise all available resources to ensure the country's readiness to respond to various threats, whether military or non-military (Fabian, 2020; Fiala, 2020, p. 3).

As part of the Comprehensive Defence Strategy (Total Defence), the civilians have a significant role where a key objective is establishing a

Table 2.2 Constitution of Ukraine

Legislation	Provision
Constitution of Ukraine, Article 17 (part)	Protecting the sovereignty and territorial integrity of Ukraine, ensuring its economic and information security, shall be the most important function of the State and a matter of concern for all the Ukrainian people

Source: (Verkhovna Rada, 1991).

civil defence to protect the population and safeguard the continuity of public services from conflict harm. This strategic step should also assist the operational capacities of the armed forces (Fiala, 2020, p. 3). Effective civil-military relations enable coordinated action and decision making between military and civilian authorities, allowing for a more effective and efficient response to crises or threats. In Ukraine, civil-military relations have been strengthened through various measures, such as establishing the National Security and Defence Council, including governmental, civilian and military officials (NSDCU, 2023). Additionally, the military has supported civilian authorities during times of crisis, such as assisting with disaster relief efforts. Civil-military relations are important to identify and coherently deal with military and security challenges, i.e., strategical, tactical and operational (Andersen & McDonald Snookermany, 2020, p. 132). To protect a state against potential threats, NATO's Comprehensive Defence Strategy Concept is an official governmental strategy incorporating the "whole-of-government" and "whole-of-society" approach. This means addressing the events that a national state and its allies/ partners consider threatening peace, security and living (Fiala, 2020, pp. 1, 5; NATO, 2020).

Resistance Operating Concept (ROC)

Wartime information and communications can gather support and influence decisions made by governments, NGOs, and influential individuals to impact public opinion nationally or internationally. Depending on the type of communication being circulated, it can boost or lower the morale of the military personnel or civilians caught up or engaged in the war (Von Tunzelmann, 2022). According to US and European officials, Ukraine has successfully used a military method of resistance warfare. This model, the Resistance Operating Concept (ROC), has been developed by US Special Operations in the aftermath of Russia's war against Georgia and the vulnerable exposure of the three Baltic NATO member states. This concept was further advanced after the 2014 attacks on the Donetsk and Luhansk regions (Fiala, 2022; Liebermann, 2022). Ukraine's ROC exemplifies how military tactics and strategies can evolve and adapt to changing geopolitical realities.

The ROC emphasises the importance of resilience and preparedness in peacetime, which can then be leveraged to resist aggression in times of war (Fiala, 2020, p. 1; Liebermann, 2022). One of the key strengths of the ROC is its flexibility and adaptability. The concept is not based on a fixed set of plans but is tailored to the specific needs of each country based on factors such as terrain, civilian abilities, and government agencies. This allows countries to develop a unique approach to resistance warfare better suited to their circumstances (Fiala, 2020, p. 7; Liebermann, 2022). It is also worth noting that the ROC involves a broader range of actors beyond the military alone. Governmental and civilian actors and agencies are included in the planning and execution of the concept. This reflects a growing recognition that modern warfare is not limited to traditional military conflicts but can involve non-state actors and asymmetric tactics (Fiala, 2020, p. 16). Overall, Ukraine's use of the ROC highlights the importance of developing innovative and adaptable military concepts to address evolving security challenges.

Resilience

Resilience in peacetime is essential to prepare for national crises and defend the state against potential threats. An aspect of this resilience is developing and planning for scenarios that could impact the nation's critical infrastructure, including cyber- and physical attacks. Inbuilt in the concept of resilience is a social and psychological defence level, which involves preparing the population to withstand and respond to potential threats. By being prepared in advance, the population can respond more effectively to threats based on their shared interests in protecting the sovereignty of their state (Fiala, 2020, p. 19; NATO, 2022b). This preparation and deterrence can help ensure that military and civil defence authorities can establish and maintain the necessary capabilities to conduct resistance activities within the national territory during a war. Overall, resilience is crucial for ensuring a nation can withstand and respond to potential crises and threats. By developing and planning for different scenarios, establishing strong civil defence systems, and preparing the population socially and psychologically, a nation can better protect itself and preserve its sovereignty (Fiala, 2019, p. 19; Roepke & Thankey, 2019).

Information warfare and the ability to collect and disseminate propaganda and disinformation has become significant challenge in modern warfare. Traditional military structures may not be sufficient to address this type of threat, requiring a more innovative defence approach. To succeed against Russian aggressions and other asymmetric threats, it may be necessary to break away from conventional military culture and incorporate irregular defence forces. These new forces could be better equipped to manage resistance operations and respond to hybrid warfare tactics (Fiala, 2020, p. 13). In addition, the online environment and cyberwarfare require innovative structures that merge military and civilian actors, focusing on developing skills and capabilities in

Table 2.3 Key ROC Definitions

Concept	Definition
Resilience	The will and ability to withstand external pressure and influence and/or recover from the effects of those pressures or influences
Resistance	A nation's organised, whole-of-society effort, encompassing the full range of activities from nonviolent to violent, led by a legally established government (potentially exiled/displaced or shown) to re-establish independence and autonomy within its sovereign territory that has been wholly or partially occupied by a foreign power

Source: (Fiala, 2019, p. 17 ; Fiala, 2020, p. xv; Maskaliūnaitė, 2021).

areas such as communications, organisation, computing, and cybersecurity. By prioritising these skills, countries can better defend themselves against the evolving threats of modern warfare (Fabian, 2020).

Resistance

On 29 July 2021, the Law of Ukraine on the Fundamentals of National Resistance legislation was signed by the Ukrainian President Zelenskyy to prepare the nation for potential aggressions from Russia. Article 1(3) states that the law includes temporarily occupied territories (Global Security, 2021; Verkhovna Rada, 2021; Fiala, 2022). The background for the legislation derives from the announcement that the Kremlin wanted to add around 20 units and formations to the western military district. This Russian move would increase the number of military forces near the border with Ukraine.

The Law of Ukraine on the Fundamentals of National Resistance aligns Ukraine's defence strategy with existing NATO practices. This legislation should be interpreted in conjunction with Ukraine's military Security Strategy, adopted in March 2021 (USCC, 2021). The Security Strategy highlights the importance of developing a territorial defence force, modernising armaments, and taking asymmetric defence actions to counter potential threats (Fiala, 2020, p. 1). It also incorporates a comprehensive security concept that includes military and non-military activities (PISM, 2021). By developing a broad concept that includes both military and non-military activities, Ukraine can better position itself to respond to the changing nature of modern warfare. In this context, the ROC can play a critical role in identifying resistance principles, requirements, and challenges that could impact doctrine, plans, capabilities, and rules of engagement development (Fiala, 2020; USCC, 2021).

Readiness to Fight

The military reform that started in 2014 decentralised the command line and transferred tactical decisions to commanders who were best equipped to make them. This means that while Zelenskyy is the Supreme Commander-in-Chief, he is not directly involved in micro-managing the forces He is supported by the General Staff of the Ukrainian Armed Forces, i.e., the Minister of Defence of Ukraine, the Commander-in-Chief of the Armed Forces of Ukraine, and the Chief of the General Staff. (Zagoronyuk et al., 2021; Onuch & Hale, 2022, p. 244). National security should be positioned at the centre of this effort, focusing on maintaining an armed force with high readiness capabilities to deter aggressors, defend the nation, secure military success, and restore national society and independence through comprehensive security and military strategies. However, the civil population also have a role in this defence structure. Pre-planning as a part of resilience is essential in managing risks in peacetime and shifting to military and civic resistance if there is an existential threat to the state's sovereignty. This requires cooperation between internal civil partners and different external partners, including military and civil actors (Fiala, 2020, pp. 3–5; NATO, 2022b). By adopting this approach, Ukraine can better prepare itself to respond to potential threats and challenges in the future.

President Putin's Threats

One problem in the conflict is directed towards the NATO. President Putin argued that the invasion was an act of self-defence against NATO's plans of enlargement, and a potential Ukrainian NATO membership poses a significant security risk to Russia. Russia also claimed that NATO is threatening Russian sovereignty and interests. Additionally, President Putin has wrongly claimed that NATO promised not to expand its organisation eastwards and include the former Soviet Union republics. However, no promises were made in 1991 when the Warsaw Pact ceased to exist (Lough, 2021; Hall, 2022; Vlamis, 2022; Wintour, 2022a). When announcing the military operation where the Russian forces entered Ukraine, President Putin warned NATO against interfering (see Table 2.4).

With the fear of escalation and triggering World War III, NATO, the EU, and other Western countries have refrained from interfering in the conflict beyond imposing economic and travel sanctions and providing weapons, money, and humanitarian aid inside and outside Ukraine (BBC News, 2022; Debusmann Jr, 2022; European Commission, 2022; Foy & Bott, 2022). There

Table 2.4 President Putin's Warning to NATO

Speaker	Quote
President Putin's speech on 24 February 2022	I would now like to say something significant for those who may be tempted to interfere in these developments from the outside. No matter who tries to stand in our way or all the more so create threats for our country and our people, they must know that Russia will respond immediately, and the consequences will be such as you have never seen in your entire history

Source: (Fisher, 2022; Lister et al., 2022; President of Russia, 2022).

has been a reluctance to provide long-range weapons to Ukraine, fearing the country would strike internally in Russia and provoke Russia to escalate its attacks. However, partner states are changing their attitude due to Russia's constant shelling of civilians and critical infrastructure in Ukraine (Brown et al., 2023). Especially Poland and the Baltic states have advocated for more weapons, including long-range missiles, to be sent to Ukraine (Euromaidan Press, 2022; Ward & Berg, 2022). The pressure from these states and voluntary civic groups on NATO member states to send tanks to Ukraine has been successful. Campaigns such as #FreeTheLeopards on Twitter have played an instrumental role in mounting pressure to embarrass states into sending tanks to Ukraine (Gressel et al., 2022; Mac Dougall, 2023).

The 2022 invasion has threatened peace and stability in Europe, and the EU member states have systematically cut their ties with Russia. The EU's unity to isolate Russia has demonstrated that the union has become a geopolitical actor. The Russian economic model is under pressure, where sanctions and the decline in economic relations have an impact (EEAS, 2022b; Meister, 2022). The 2022 G20 summit in Bali has also displayed Russia's isolation. Most G20 nations strongly condemned the war and repeated earlier demands that Russia should withdraw from the Ukrainian territory. Even China and India, which had previously stepped back from critiquing Russia publicly, showed a significant lack of support for Russia (Economist Intelligence, 2022). China's President Xi warned against weaponising food and energy and opposed using nuclear weapons/ nuclear war under all circumstances. India's Prime Minister Modi called for a ceasefire and diplomacy to solve the situation in Ukraine (Wintour, 2022b). The G20 joint statement condemning Russia's war in Ukraine claimed: "Today's era must not be of war" (White House, 2022).

Tensions and False Justifications

One People

Russia has long pursued the narrative that Ukraine is not a state but an integrated part of Russia. This line of argument has worked internally in Russia to justify the annexation of Crimea and the surrounding regions in the southwestern part of Ukraine.

Table 2.5 President Putin's Speech at the 2008 NATO Summit

Speaker	Quote
President Putin's speech in 2008	Ukraine is not even a state! What is Ukraine? A part of its territory is [in] Eastern Europe, but a[nother] part, a considerable one, was a gift from us!

Source: (Arnold, 2020; U.S. Department of State, 2022).

Table 2.6 President Putin's Essay about Historical Unity

Author	Quotes
President Putin's essay from 12 July 2021	Russians, Ukrainians, and Belarusians are all descendants of Ancient Rus, which was the largest state in Europe. Slavic and other tribes across the vast territory – from Ladoga, Novgorod, and Pskov to Kiev and Chernigov – were bound together by one language (which we now refer to as Old Russian), economic ties, the rule of the princes of the Rurik dynasty, and – after the baptism of Rus – the Orthodox faith. The spiritual choice made by St. Vladimir, who was both Prince of Novgorod and Grand Prince of Kiev, still largely determines our affinity today

Source: (President of Russia, 2021).

After the illegal annexation of Crimea in 2014, President Putin continued to propagate this narrative. In 2020, Putin reiterated the mantra that Russians and Ukrainians are one people, an argument that the Kremlin has often used to justify its actions. Putin went further in his 2021 essay On the Historical Unity between Russians and Ukrainians, in which he promoted the idea of a pseudo-historical unity between the two nations based on a mixture of dubious claims and grievances, emphasising the notion of "blood ties" (President of Russia, 2021; Harding, 2022, p. 23; Harris et al., 2022). The assumption that Ukraine's national identity is a product of foreign interference is also a common theme in Putin's rhetoric (Arnold, 2020).

These arguments and proof for historical and cultural belonging forwarded in the article have several shortfalls. First, Putin separated the three populations as Russians, Ukrainians, and Belarusians. Later, he linked them together as if all three populations were Russians (President of Russia, 2021; Wilson, 2021). Second, the argument that Ukraine is an integral part of Russia is not supported by historical or legal evidence. Putin largely ignores that Ukraine is a sovereign state, just as Russia has sovereignty. Sovereignty means Ukraine can decide independently about matters concerning the country without interference (Verkhovna Rada, 1990; Wilson, 2021). Third, Putin seems to believe a common language and religion exist. He rejects the existence of a Ukrainian language and that the Ukrainian Orthodox Church broke ties with the Moscow Patriarchate (President of Russia, 2021; Wilson, 2021; Chotiner, 2022; Hayda, 2022; Reuters, 2022b). Last, Putin's narrative ignores Ukraine's rich history and culture, including the medieval Kyivan Rus, which is considered the precursor to Russian and Ukrainian culture (Andrejsons, 2022; Schwirtz et al., 2022). In this context, Putin also declared that: "[T]he true sovereignty of Ukraine is possible only in partnership with Russia" (U.S. Department of State, 2022).

To any country other than Russia, the actions to demolish another state its independence is an imperial raid to rein in a former dependent state (Harding, 2022, p. 6).

In his address on 21 February 2022, President Putin's recognition of the independence and sovereignty of the self-proclaimed "DPR" and "LPR" clearly violated Ukraine's territorial integrity and sovereignty (President of Russia, 2022). President Putin is essentially acknowledging the two states' claims to be separate from Ukraine and to have the right to self-determination, a move that will undermine Ukraine's control over its territory. This recognition contradicts his earlier claims that Ukraine and Russia are one country, as the statement implies that the two states are separate political entities with their own sovereignty.

In conclusion, Putin's arguments for Ukraine's historical and cultural belonging to Russia lack substantial evidence and ignore Ukraine's sovereignty as a state (see Table 2.7). These arguments serve to justify Russia's aggression towards Ukraine and the annexation of Crimea. President Putin's arguments for the enlargement of Russia stand in sharp contrast to Ukraine's President Zelenskyy's speech on the celebration of the Day of Ukrainian Statehood in 2022 after 155 days of fighting (see Table 2.8).

Ukrainians are actively engaged in the war, defending their country and their right to exist as an independent nation. The Kremlin's attempts to eradicate Ukraine as a sovereign state have only strengthened the population's determination to resist and refuse to be seen as "little Russians" (Dyczok, 2022).

Table 2.7 President Putin's Vision of "One Country"

Speaker	Quote
President Putin's Speech on 21 February 2022	I would like to emphasise again that Ukraine is not just a neighbouring country for us. It is an inalienable part of our own history, culture and spiritual space. These are our comrades, those dearest to us – not only colleagues, friends and people who once served together, but also relatives, people bound by blood, by family ties

Source: (President of Russia, 2022).

Table 2.8 President Zelenskyy's Rejection of Ukraine and Russia as "One Country"

Speaker	Quote
President Zelenskyy's speech on 28 July 2022	Every day we fight so that everyone on the planet finally understands: we are not a colony, not an enclave, not a protectorate. Not a gubernia, an eyalet or a crown land, not a part of foreign empires, not a 'part of the land', not a union republic. Not an autonomy, not a province, but a free, independent, sovereign, indivisible and independent state

Source: (President of Ukraine, 2022; U.S. Department of State, 2022).

De-Nazify Ukraine

The Kremlin's defence for launching its so-called "special military operation" in 2022 is to "de-militarise" and "de-Nazify" Ukraine, ending eight years of bullying and genocide against Russians living in the country (Fisher, 2022; Kirby, 2022a; Lister et al., 2022; Veidlinger, 2022). In his speech on 21 February 2022, President Putin continued to argue that the authorities in Kyiv had taken an aggressive stance against Russia because of the annexation. Therefore, the Ukrainian government has activated extremist cells and suppressed everything related to Russian culture, language, and religion (see Table 2.9) (President of Russia, 2022).

The "neo-Nazi" argument can be traced back to the Euromaidan protest, where President Putin constantly pursued the idea that radical nationalists had taken control of Ukraine (President of Russia, 2022). The Russian state-controlled mass media outlets have advanced the view that Euromaidan was a coup conducted by Ukrainian neo-Nazis and foreign agents. However, in the narrative advanced by Russia, neo-Nazi stereotyping is still being used to explain the different aggressions against Ukraine. Although individual nationalist groups took part in the Euromaidan revolution in 2013–2014, these groups did not have a prominent position; instead, they had a subordinate role during the events. It was not a protest favouring developing an ethnonational state (Yekelchyk, 2020, pp. 95, 97).

Russian and western critics have also repeated the argument regarding far-right groups to support the idea that the new government was illegal and was seen as a threat to the Russian minority in Ukraine (D'Anieri, 2019, p. 222). After the Euromaidan Revolution, this part of the movement with ties to the radical right transformed into a tiny fraction within the political environment.

Table 2.9 President Putin's Neo-Nazi Speech

Speaker	Quote
President Putin's speech on 21 February 2022	At the same time, the Ukrainian authorities – I would like to emphasise this – began by building their statehood on the negation of everything that united us, trying to distort the mentality and historical memory of millions of people, of entire generations living in Ukraine. It is not surprising that Ukrainian society was faced with the rise of far-right nationalism, which rapidly developed into aggressive Russophobia and neo-Nazism. This resulted in the participation of Ukrainian nationalists and neo-Nazis in the terrorist groups in the North Caucasus and the increasingly loud territorial claims to Russia

Source: (President of Russia, 2022).

The link became hyped at the beginning of the war in the Donetsk and Luhansk regions, where it again served Russian interests to sow social divisions by arguing that far-right extremists fought the war. Indeed, some volunteer battalions involved in the fight were formed by far-right activists (Yekelchyk, 2020, p. 97). The Ukrainian Azov Regiment has been criticised for its right-wing links. Yet, it was separated from the movements in 2014 when it was incorporated into the Interior Ministry's troops, the National Guard (Yekelchyk, 2020, p. 97; Weber, 2022).

Russia's attempts to portray Ukraine as a hotbed of neo-Nazism are also contradicted by reality. While there may be individual instances of extremist behaviour, Ukraine is a multi-ethnic and multicultural society that has taken steps to promote diversity and tolerance (UN, 2016; UNESCO, 2019; CoE, 2022). The Kremlin's use of the neo-Nazi argument to justify its aggression towards Ukraine is based on distorted facts and furthers Russian interests by creating divisions and justifying military action. The use of such arguments is a reflection of Russia's propaganda efforts to manipulate international opinion and obscure the true nature of its actions (Kirby, 2022b). The Nazi arguments are often used to influence the historical memory of states turning towards the West. Russia is trying to highlight how countries such as Ukraine and the Baltic states suffered under the German Nazi regime during the Great Patriotic War (WWII) and how they were saved by the Red Army (Thornton, 2015, p. 43). However, it is paradoxical that Russian war strategies, destruction, torture, and attacks on civilians and critical infrastructure mirror those actions conducted by Nazi Germany (Guerin, 2022; HRW, 2022; Khurshudyan & Robinson Chavez, 2022).

President Putin tried again to use the same Nazi argument in his speech on 24 February 2022. The purpose of the "special military operation" was to protect people by claiming that the Russian-speaking or ethnic Russians in Ukraine had been:

[S]ubjected to bullying and genocide … for the last eight years. And for this we will strive for the demilitarisation and denazification of Ukraine. (Veidlinger, 2022)

The claim that Russia aims to de-Nazify Ukraine, especially the Ukrainian government, is also inconsistent since Ukraine's Jewish President Zelenskyy's family members fought against Nazi Germany on the Soviet side (Sheftalovich, 2022; Sonne, 2022; Veidlinger, 2022). In a televised speech directed at the Russian public just before President Putin announced his "special military operation" and subsequent invasion of Ukraine, President Zelenskyy delivered an emotional plea for peace in Russia, which, unfortunately, proved unsuccessful (see Table 2.10).

Table 2.10 President Zelenskyy's Plea

Speaker	Quote
President Zelenskyy's speech on 23 February 2022	The Ukraine on your news and Ukraine in real life are two completely different countries – and the main difference between them is: Ours is real. You are told we are Nazis. But could a people who lost more than 8 million lives in the battle against Nazism support Nazism?

Source: (Sheftalovich, 2022; Sonne, 2022).

Nationalism and Resilience

Ukraine's resilience and successful defence are linked not only to the military force but also to the Ukrainian people's resilience against aggression from Russia, which before the invasion, was considered a military superpower on multiple levels, offline and online, urban and rural.

The level of defence and resilience displayed by the Ukrainians has earned admiration from the rest of the world (CRS, 2022, p. 2; Harari, 2022). Ukraine's "whole-of-government" and "whole-of-society" approach has involved various activities, including both violent and nonviolent approaches, that a legally established government has initiated. This achievement resulted from planning and coordination across different military, government, and societal strategies (Fiala, 2020, p. 5).

However, the Kremlin failed to recognise that Ukraine had mobilised and changed since Russia illegally annexed Crimea in 2014, and that the military is now more strategically and tactically prepared for the 2022 war. After Ukraine gained independence, the country struggled to develop a unified national identity that embraced ethnic, religious, and linguistic differences. Ukraine has transformed from a largely fragmented state to obtaining unity around a national state independent of ethnonationalism (Lohsen, 2022). Attention has been given to the rise of nationalism in Ukraine, in spite of its being less noticeable than in Russia (Kuzio & D'Anieri, 2018, p. 15). Before the annexation of Crimea, none of the extreme nationalist parties managed to get enough votes in the parliament, and there were no electoral bases for promoting an ethnic nationalist program for a potential presidential candidate (Kuzio & D'Anieri, 2018, p. 15).

There has been a notable increase in the Ukrainian national identity across different regions, including those traditionally more pro-Russian. This has affected the meaning of "being Ukrainian", closely tied to the Ukrainian state. On social media, posts and videos from Ukrainian volunteers and military personnel in the war zone show individuals who are both Russian and Ukrainian speakers, indicating that they are fighting for a united Ukraine regardless of their ethnic background (Yekelchyk, 2020, p. 19; Barrington, 2022; Epstein & David, 2022; Melkozerova, 2022). The concepts of resilience and resistance in ROC feed into the idea of the national state, where resilience is the will of the people to maintain the sovereignty that they have obtained, and resistance is a natural response to an existential threat to the power and independence of the state (Fiala, 2020, p. 5). This change has been visible since 2014, when national and cultural identity emerged beyond ethnicity, including Ukrainian citizenship. Ukraine's strong national and cultural identity withstood the Russian myth of an all-Russian nation and the use of "one nation" terminology to legitimise Russian aggressions (EEAS, 2022a). The war has prompted a reconsideration of the Russian and Ukrainian shared history, as Ukrainians reject the Russian ideology of "one nation/one people". From Ukraine's viewpoint, the shared history is based not on a fraternal nation but on a history of colonial power and suppression. The two nations' views on the past, present, and future are fundamentally incompatible (Lohsen, 2022).

References

Allan, D., 2020. *The Minsk conundrum: Western policy and Russia's war in eastern Ukraine.* [Online] Available at: https://www.chathamhouse.org/2020/05/minsk-conundrum-western-policy-and-russias-war-eastern-ukraine-0/minsk-2-agreement [Accessed 17 03 2022].

Andersen, M. & McDonald Snookermany, A., 2020. The Making of Military Strategy. The Gravity of an Unequal Dialogue. In: *Military Strategy in the 21st Century. The Challenge for NATO.* London: Hurst, pp. 131–151.

Andrejsons, K., 2022. *Russia and Ukraine are trapped in medieval myths.* [Online] Available at: https://foreignpolicy.com/2022/02/06/russia-and-ukraine-are-trapped-in-medieval-myths/ [Accessed 28 03 2023].

AP, 2022. *EXPLAINER: What's behind referendums in occupied Ukraine?* [Online] Available at: https://apnews.com/article/russia-ukraine-moscow-referendums-crimea-dfa7e0aa150b2e12bf92ac070ba00879 [Accessed 03 01 2023].

Arnold, K., 2020. *"There is no Ukraine": Fact-checking the Kremlin's version of Ukrainian history.* [Online] Available at: https://blogs.lse.ac.uk/lseih/ 2020/07/01/there-is-no-ukraine-fact-checking-the-kremlins-version-of-ukrainian-history/ [Accessed 17 12 2022].

Arnold, E., 2022. *Slava Ukraini: Assessing the Ukrainian will to fight.* [Online] Available at: https://rusi.org/explore-our-research/publications/commentary/slava-ukraini-assessing-ukrainian-will-fight [Accessed 18 12 2022].

Atlantic Council, 2014. *A roadmap for Ukraine: Delivering on the promise of the maidan,* s.l.: Jstor.

Barrington, L., 2022. *Putin's key mistake? Not understanding Ukraine's blossoming national identity – even in the Russian-friendly southeast.* [Online] Available at: https://theconversation.com/putins-key-mistake-not-understanding-ukraines-blossoming-national-identity-even-in-the-russian-friendly-southeast-183576 [Accessed 18 09 2022].

BBC News, 2014a. *Crimea referendum: Voters "back Russia union".* [Online] Available at: https://www.bbc.co.uk/news/world-europe-26606097 [Accessed 03 01 2023].

BBC News, 2014b. *Profile: Ukraine's ousted President Viktor Yanukovych.* [Online] Available at: https://www.bbc.co.uk/news/world-europe-25182830 [Accessed 23 03 2023].

BBC News, 2015. *Ukraine ceasefire: New Minsk agreement key points.* [Online] Available at: https://www.bbc.co.uk/news/world-europe-31436513 [Accessed 03 01 2023].

BBC News, 2016. *Mikhail Gorbachev: The man who lost an empire.* [Online] Available at: https://www.bbc.co.uk/news/world-europe-38289333 [Accessed 03 01 2023

BBC News, 2022. *What sanctions are being imposed on Russia over Ukraine invasion?* [Online] Available at: https://www.bbc.co.uk/news/world-europe-60125659 [Accessed 21 03 2022].

Bonenberger, A., 2022. *Ukraine's military pulled itself out of the ruins of 2014.* [Online] Available at: https://foreignpolicy.com/2022/05/09/ukraine-military-2014-russia-us-training/ [Accessed 24 12 2022].

Booth, W., 2014. *Ukraine's parliament votes to oust president; former prime minister is freed from prison.* [Online] Available at: https://www.washingtonpost.com/world/europe/ukraines-yanukovych-missing-as-protesters-take-control-of-presidential-residence-in-kiev/2014/02/22/802f7c6c-9bd2-11e3-ad71-e03637a299c0_story.html [Accessed 23 03 2023].

Borda, A., 2022. *Ukraine war: What is the Budapest Memorandum and why has Russia's invasion torn it up?* [Online] Available at: https://theconversation.com/ukraine-war-what-is-the-budapest-memorandum-and-why-has-russias-invasion-torn-it-up-178184 [Accessed 20 03 2023].

Brown, J., Horton, J., & Ahmedzade, T., 2023. *Ukraine weapons: What military equipment is the world giving?* [Online] Available at: https://www.bbc.co.uk/news/world-europe-62002218 [Accessed 15 01 2023].

Brzozowski, A., 2022. *Zelenskyy blames Germany, France over failed Ukraine diplomacy.* [Online] Available at: https://www.euractiv.com/section/europe-s-east/news/zelenskyy-blames-germany-france-over-failed-ukraine-diplomacy/[Accessed 24 12 2022].

Chappell, B. & Memmott, M., 2014. *Putin says those aren't Russian forces in Crimea.* [Online] Available at: https://www.npr.org/sections/thetwo-way/2014/03/04/ 285653335/putin-says-those-arent-russian-forces-in-crimea [Accessed 21 12 2022].

Chotiner, I., 2022. *Vladimir Putin's revisionist history of Russia and Ukraine.* [Online] Available at: https://www.newyorker.com/news/q-and-a/vladimir-putins-revisionist-history-of-russia-and-ukraine [Accessed 15 01 2023].

CoE, 2014. *The illegal annexation of Crimea has no legal effect and is not recognised by the Council of Europe.* [Online] Available at: https://pace.coe.int/en/news/4975/the-illegal-annexation-of-crimea-has-no-legal-effect-and-is-not-recognised-by-the-council-of-europe [Accessed 03 01 2023].

CoE, 2022. *Fifth Report submitted by Ukraine.* Strasburg: ACFC.

Commander-in-Chief of the Armed Forces of Ukraine, 2023. *@CinC_AFU.* [Online] Available at: https://mobile.twitter.com/CinC_AFU/status/1609948188550598659 [Accessed 03 01 2023].

Crisis Group, 2022. *Conflict in Ukraine's Donbas: A visual explainer.* [Online] Available at: https://www.crisisgroup.org/content/conflict-ukraines-donbas-visual-explainer [Accessed 22 12 2022].

CRS, 2022. *Russia's war in Ukraine: Military and intelligence aspects,* Washington, DC: Congressional Research Service.

D'Anieri, P., 2019. *Ukraine and Russia. From Civilised Divorce to Uncivil War.* Cambridge: Cambridge University Press.

Debusmann Jr, B., 2022. *What weapons will US give Ukraine – and how much will they help?*. [Online] Available at: https://www.bbc.co.uk/news/world-us-canada-60774098 [Accessed 18 03 2022].

Dickinson, P., 2020. *How Ukraine's Orange Revolution shaped twenty-first century geopolitics.* [Online] Available at: https://www.atlanticcouncil.org/blogs/ ukrainealert/ how-ukraines-orange-revolution-shaped-twenty-first-century-geopolitics/[Accessed 17 12 2021].

Dickinson, P., 2022. *How modern Ukraine was made on maidan.* [Online] Available at: https://www.atlanticcouncil.org/blogs/ukrainealert/how-modern-ukraine-was-made-on-maidan/ [Accessed 12 17 2022].

Dyczok, M., 2022. *A former journalist recalls Ukraine's 1991 vote for independence — and how its resilience endures.* [Online] Available at: https://theconversation. com/a-former-journalist-recalls-ukraines-1991-vote-for-independence-and-how-its-resilience-endures-189266 [Accessed 17 08 2022].

Economist Intelligence, 2022. *G20 summit indicates Russia's growing isolation globally.* [Online] Available at: https://www.eiu.com/n/g20-summit-indicates-russia-growing-isolation-globally/ [Accessed 24 12 2022].

EEAS, 2022a. *Disinformation about Russia's invasion of Ukraine – debunking seven myths spread by Russia.* [Online] Available at: https://www.eeas.europa.eu/ delegations/ china/disinformation-about-russias-invasion-ukraine-debunking-seven-myths-spread-russia_en?s=166 [Accessed 11 11 2022].

EEAS, 2022b. *Putin's war has given birth to geopolitical Europe.* [Online] Available at: https://www.eeas.europa.eu/eeas/putins-war-has-given-birth-geopolitical-europe_en [Accessed 24 12 2022].

Epstein, J. & David, C.R., 2022. *Putin thought Russia's military could capture Kyiv in 2 days, but it still hasn't in 20.* [Online] Available at: https://www.businessinsider.com/ vladimir-putin-russian-forces-could-take-kyiv-ukraine-two-days-2022–3?r=US& IR=T [Accessed 18 09 2022].

Erlanger, S. & Myers, S.L., 2008. *Nato allies oppose Bush on Georgia and Ukraine.* [Online] Available at: https://www.nytimes.com/2008/04/03/ world/europe/03nato. html [Accessed 24 12 2022].

Euromaidan Press, 2022. *UK can send long-range weapons to Ukraine – British Defence Secretary.* [Online] Available at: https://euromaidanpress.com/2022/12/13/ uk-can-send-long-range-weapons-to-ukraine-british-defence-secretary/ [Accessed 15 01 2023].

European Commission, 2022. *Ukraine: EU agrees fourth package of restrictive measures against Russia.* [Online] Available at: https://ec.europa.eu/commission/ presscorner/ detail/en/ip_22_1761 [Accessed 21 03 2022].

European Parliament, 2016. *Ukraine and the Minsk II agreement. On a frozen path to peace?* [Online] Available at: https://www.europarl.europa.eu/RegData/etudes/ BRIE/2016/ 573951/EPRS_BRI(2016)573951_EN.pdf [Accessed 17 03 2022].

Fabian, S., 2020. *Total defense, revisited: An unconventional solution to the problem of conventional force.* [Online] Available at: https://mwi.usma.edu/total-defense-revisited-unconventional-solution-problem-conventional-forces/ [Accessed 15 01 2023].

Fiala, O., 2019. *Resistance Operating Concept.* Stockholm: Special Operations Command Europe (SOCEUR) & the Swedish Defence University.

Fiala, O.C., 2020. *ROC. Resistance Operating Concept.* MacDill Air Force Base(Florida): JSOU Press.

Fiala, O., 2022. Resilience and Resistance in Ukraine. *Small Wars Journal*, 31, p. 12.

Fisher, M., 2022. *Putin's case for war, annotated.* [Online] Available at: https://www.nytimes.com/2022/02/24/world/europe/putin-ukraine-speech.html [Accessed 20 03 2022].

Foy, H. & Bott, I., 2022. *How is Ukraine using western weapons to exploit Russian weaknesses?* [Online] Available at: https://www.ft.com/content/f5fb2996-f816-4011-a440-30350fa48831 [Accessed 21 03 2022].

France24, 2022. *Merkel defends 2008 decision to block Ukraine from NATO.* [Online] Available at: https://www.france24.com/en/live-news/20220404-merkel-defends-2008-decision-to-block-ukraine-from-nato [Accessed 24 12 2022].

Global Conflict Tracker, 2022. *Conflict in Ukraine.* [Online] Available at: https://www.cfr.org/global-conflict-tracker/conflict/conflict-ukraine [Accessed 03 01 2022].

Global Security, 2021. *Territorial defense forces TrO – doctrine.* [Online] Available at: https://www.globalsecurity.org/military/world/ukraine/tro-doctrine.htm [Accessed 02 01 2022].

Gorenburg, D., 2015. Editor's Introduction. *Russian Politics and Law*, 53(3), pp. 1–5.

GOV.UA, 2020. *National defence.* [Online] Available at: https://www.kmu.gov.ua/en/reformi/bezpeka-ta-oborona/national-defence [Accessed 24 12 2022].

Gressel, G., Loss, R., & Puglierin, 2022. *The Leopard plan: How European tanks can help Ukraine take back its territory.* [Online] Available at: https://ecfr.eu/article/the-leopard-plan-how-european-tanks-can-help-ukraine-take-back-its-territory/ [Accessed 15 01 2023].

Grytsenko, O., 2014. *Pro-Russia groups take over government buildings across Ukraine.* [Online] Available at: https://www.theguardian.com/world/2014/mar/03/pro-russia-groups-government-buildings-ukraine [Accessed 21 12 2022].

Guerin, O., 2022. *Ukraine war: Accounts of Russian torture emerge in liberated areas.* [Online] Available at: https://www.bbc.co.uk/news/world-europe-62888388 [Accessed 19 09 2022].

Haaland Matlary, J. & Johnson, R., 2020. Introduction. In: *Military Strategy in the 21st Century. The Challenge for NATO.* London: Hurst, pp. 1–26.

Hall, G.E.L., 2022. *Ukraine: The history behind Russia's claim that Nato promised not to expand to the east.* [Online] Available at: https://theconversation.com/ukraine-the-history-behind-russias-claim-that-nato-promised-not-to-expand-to-the-east-177085 [Accessed 17 03 2022].

Harari, Y.N., 2022. *Why Vladimir Putin has already lost this war.* [Online] Available at: https://www.theguardian.com/commentisfree/2022/feb/28/vladimir-putin-war-russia-ukraine [Accessed 18 09 2022].

Harding, L., 2022. *Demoralised Russian soldiers tell of anger at being 'duped' into war.* [Online] Available at: https://www.theguardian.com/world/2022/mar/04/russian-soldiers-ukraine-anger-duped-into-war [Accessed 19 09 2022].

Harris, S., De Young, K., Khurshudyan, I., Parker, A., Sly, L. 2022. *Road to war: U.S. struggled to convince allies, and Zelensky, of risk of invasion.* [Online] Available at: https://www.washingtonpost.com/national-security/interactive/2022/ukraine-road-to-war/?itid=lb_war-in-ukraine-what-you-need-to-know_3 [Accessed 23 12 2022].

Hayda, J., 2022. *Orthodox Church in Ukraine has decided to cut ties with Russia.* [Online] Available at: https://www.npr.org/2022/05/28/1101921353/orthodox-church-in-ukraine-has-decided-to-cut-ties-with-russia [Accessed 15 01 2023].

Herszenhorn, D.M. & McLeary, P., 2022. *Ukraine's "iron general" is a hero, but he's no star.* [Online] Available at: https://www.politico.com/news/2022/04/08/ukraines-iron-general-zaluzhnyy-0002390 [Accessed 05 01 2022].

Hockaday, J., 2022. *Zelensky translator chokes up after Ukraine president says 'nobody will break us'.* [Online] Available at: https://metro.co.uk/2022/03/01/zelensky-translator-chokes-up-after-ukraine-president-says-nobody-will-break-us-16194896/ [Accessed 30 12 2022].

Horvath, R., 2011. Putin's "Preventive Counter-Revolution": Post-Soviet authoritarianism and the spectre of Velvet Revolution. *Europe-Asia Studies*, 63(1), pp. 1–25.

HRW, 2022. *Ukraine: Apparent war crimes in Russia-controlled areas.* [Online] Available at: https://www.hrw.org/news/2022/04/03/ukraine-apparent-war-crimes-russia-controlled-areas [Accessed 19 09 2022].

Hutchinson, B. & Reevell, P., 2022. *What are the Ukraine "separatist" regions at the crux of the Russian invasion?* [Online] Available at: https://abcnews.go.com/ International/ ukraine-separatist-regions-crux-russian-invasion/story?id=83084803 [Accessed 03 01 2023].

Katchanovski, I., 2008. The Orange Evolution? The "Orange Revolution" and Political Changes in Ukraine. *Post-Soviet Affairs*, 24(4), pp. 351–382.

Khurshudyan, I. & Robinson Chavez, M., 2022. *Ukrainian villagers describe cruel.* [Online] Available at: https://www.washingtonpost.com/world/interactive/2022/ russian-soldiers-beat-torture-ukrainian-villagers/ [Accessed 19 09 2022].

Kingsley, T. & Sommerlad, J., 2022. *Why has Russia invaded Ukraine? The conflict simply explained.* [Online] Available at: https://www.independent.co.uk/ news/world/europe/ why-russia-invaded-ukraine-war-explained-b2037843.html [Accessed 17 03 2022].

Kirby, P., 2022a. *What does Putin want and will Russia end its war?* [Online] Available at: https://www.bbc.co.uk/news/world-europe-56720589 [Accessed 17 03 2022].

Kirby, P., 2022b. *Why has Russia invaded Ukraine and what does Putin want?* [Online] Available at: https://www.bbc.co.uk/news/world-europe-56720589 [Accessed 18 09 2022].

Klid, B., 2007. Rock, Pop and Politics in Ukraine's 2004 Presidential Campaign and Orange Revolution. *Journal of Communist Studies and Transition Politics*, 23(1), pp. 118–137.

Krushelnycky, A., 2013. *The fight for the maidan.* [Online] Available at: https://foreign-policy.com/2013/12/13/the-fight-for-the-maidan/ [Accessed 17 12 2022].

Kulyk, V., 2016. National Identity in Ukraine: Impact of Euromaidan and the war. *Europe-Asia Studies*, 68(4), pp. 588–608.

Kuzio, T. & D'Anieri, P., 2018. *The Sources of Russia's Great Power Politics. Ukraine and the Challenge to the European Order.* 1st ed. Bristol: E-International Relations.

Lane, D., 2008. The Orange Revolution: "People's Revolution" or revolutionary coup? *British Journal of Politics and International Relations*, 10(4), pp. 525–549.

Larsson, S., 2020. *Swedish Total Defence and the Emergence of Societal Security.* Abingdon: Routledge.

Liebermann, O., 2022. *How Ukraine is using resistance warfare developed by the US to fight back against Russia.* [Online] Available at: https://edition.cnn.com/2022/08/27/ politics/russia-ukraine-resistance-warfare/index.html [Accessed 24 12 2022].

Lister, T., John, T., & Murphy, P.P., 2022. *Here's what we know about how Russia's invasion of Ukraine unfolded.* [Online] Available at: https://edition.cnn.com/2022/02/24/ europe/ukraine-russia-attack-timeline-intl/index.html [Accessed 17 03 2022].

Lohsen, A., 2022. *How the war could transform Ukrainian politics.* [Online] Available at: https://www.csis.org/analysis/how-war-could-transform-ukrainian- politics [Accessed 13 11 2022].

Lough, J., 2021. *Myth 03: "Russia Was Promised that NATO Would Not Enlarge".* London: Chatham House.

Luhn, A. & Grytsenko, O., 2015. *Ukrainian soldiers share horrors of Debaltseve battle after stinging defeat.* [Online] Available at: https://www.theguardian.com/world/2015/ feb/18/ukrainian-soldiers-share-horrors-of-debaltseve-battle-after-stinging-defeat [Accessed 15 01 2023].

Mac Dougall, D., 2023. *"Free the Leopards!" Campaign aims to "embarrass" Germany into sending tanks to Ukraine.* [Online] Available at: https://www.euronews.com/2023/01/05/free-the-leopards-campaign-aims-to-embarrass-germany-into-sending-tanks-to-ukraine [Accessed 15 01 2023].

Maskaliūnaitė, A., 2021. Exploring Resistance Operating Concept. Promises and Pitfalls of (Violent) Underground Resistance. *Journal on Baltic Security*, 7(1), pp. 27–38.

Meister, S., 2022. *A paradigm shift: EU–Russia relations after the war in Ukraine.* [Online] Available at: https://carnegieeurope.eu/2022/11/29/paradigm-shift-eu-russia-relations-after-war-in-ukraine-pub-88476 [Accessed 24 12 2022].

Melkozerova, V., 2022. *The war tat Russia already lost.* [Online] Available at: https://www.theatlantic.com/ideas/archive/2022/10/russia-war-ukraine-national-identity/671685/ [Accessed 11 11 2022].

Munk, T., 2022. *The Rise of Politically Motivated Cyber Attacks.* London: Routledge.

Myre, G., 2021. *How the Soviet Union's collapse explains the current Russia–Ukraine tension.* [Online] Available at: https://www.npr.org/2021/12/24/1066861022/how-the-soviet-unions-collapse-explains-the-current-russia-ukraine-tension [Accessed 17 12 2022].

NATO, 1949. *The North Atlantic Treaty.* [Online] Available at: https://www.nato.int/cps/en/natolive/official_texts_17120.htm [Accessed 19 03 2023].

NATO, 2008. *Bucharest Summit Declaration.* [Online] Available at: https://www.nato.int/cps/en/natolive/official_texts_8443.htm [Accessed 23 03 2023].

NATO, 2020. *Comprehensive Defence Handbook, Volume I.* Brussels: NATO Special Operations Headquarters.

NATO, 2022a. *Relations with Ukraine.* [Online] Available at: https://www.nato.int/ cps/en/natohq/topics_37750.htm [Accessed 24 12 2022].

NATO, 2022b. *Resilience, civil preparedness and Article 3.* [Online] Available at: https://www.nato.int/cps/en/natohq/topics_132722.htm [Accessed 26 02 2023].

NBCNews, 2008. *Ukraine's prime minister urges factions to unite.* [Online] Available at: https://www.nbcnews.com/id/wbna27267211 [Accessed 29 03 2023].

Noorman, R., 2020. The Battle of Debaltseve: A hybrid army in a classic battle of encirclement. *Small Wars Journal*, 17, p. 7.

NSDCU, 2023. *National security and Defence Council of Ukraine.* [Online] Available at: https://www.rnbo.gov.ua/en/ [Accessed 23 03 2023].

Onuch, O. & Hale, H.E., 2022. *The Zelensky Effect.* London: Hurst.

OSCE, 1975. *Helsinki final act.* [Online] Available at: https://www.osce.org/helsinki-final-act [Accessed 20 03 2023].

Pifer, S., 2020. *Crimea: Six years after illegal annexation.* [Online] Available at: https://www.brookings.edu/blog/order-from-chaos/2020/03/17/crimea-six-years-after-illegal-annexation/ [Accessed 17 12 2022].

PISM, 2021. *Ukraine's new military security strategy.* [Online] Available at: https://pism. pl/publications/Ukraines_New_Military_Security_Strategy [Accessed 02 01 2023].

President of Russia, 2021. *Article by Vladimir Putin "On the Historical Unity of Russians and Ukrainians".* [Online] Available at: http://en.kremlin.ru/events/president/news/66181 [Accessed 15 01 2021].

President of Russia, 2022. *Address by the President of the Russian Federation.* [Online] Available at: https://archive.ph/58LtK#selection-761.0-761.50 [Accessed 17 12 2022].

President of Ukraine, 2021. *Head of State approves strategic defense bulletin of Ukraine.* [Online] Available at: https://www.president.gov.ua/en/news/glava-derzhavi-zatverdiv-strategichnij-oboronnij-byuleten-uk-70713 [Accessed 24 12 2022].

President of Ukraine, 2022. *All stages of the history of Ukrainian statehood can be described in one sentence: We existed, exist and will exist – address by President Volodymyr Zelenskyy on the occasion of Ukrainian Statehood Day.* [Online] Available at: https://www.president.gov.ua/en/news/vsi-etapi-istoriyi-derzhavnosti-ukrayini-mozhna-opisati-odni-76705#:~:text=We%20can%20say%20that%20for, %22%2C%20not%20a%20union%20republic [Accessed 17 12 2022].

Reuters, 2007. *FACTBOX: Who is President Viktor Yushchenko?* [Online] Available at: https://www.reuters.com/article/us-ukraine-election-yushchenko-idUSL271055720070927 [Accessed 21 03 2023].

Reuters, 2022a. *Factbox: What are the Minsk agreements on the Ukraine conflict?* [Online] Available at: https://www.reuters.com/world/europe/what-are-minsk-agreements-ukraine-conflict-2022-02-21/ [Accessed 03 01 2023].

Reuters, 2022b. *Moscow-led Ukrainian Orthodox Church breaks ties with Russia.* [Online] Available at: https://www.reuters.com/world/europe/moscow-led-ukrainian-orthodox-church-breaks-ties-with-russia-2022-05-28/ [Accessed 15 01 2023].

Roepke, W. & Thankey, H., 2019. *Resilience: The first line of defence.* [Online] Available at: https://www.nato.int/docu/review/articles/2019/02/27/resilience-the-first-line-of-defence/index.html [Accessed 26 03 2023].

Roth, A., 2022. *What is the background to the separatist attack in east Ukraine?* [Online] Available at: https://www.theguardian.com/world/2022/feb/17/what-is-the-background-to-the-separatists-attack-in-east-ukraine [Accessed 03 01 2023].

Schwirtz, M., Varenikova, M., & Gladstone, R., 2022. *Putin calls Ukrainian statehood a fiction. History suggests otherwise.* [Online] Available at: https://www.nytimes. com/2022/02/21/world/europe/putin-ukraine.html [Accessed 27 03 2023].

Sheftalovich, Z., 2022. *Ukraine's Jewish President Zelenskiy asks Putin: "How could I be a Nazi?"* [Online] Available at: https://www.politico.eu/article/ukraine-jewish-president-volodymyr-zelenskiy-how-could-i-be-a-nazi-vladimir-putin-war/ [Accessed 23 12 2022].

Shelest, H., 2022. *Defend. Resist. Repeat: Ukraine's lessons for European defence.* [Online] Available at: https://ecfr.eu/publication/defend-resist-repeat-ukraines-lessons-for-european-defence/ [Accessed 02 01 2023].

Shveda, Y. & Park, J., 2016. Ukraine's revolution of dignity: The dynamics of Euromaidan. *Journal of Euroasian Studies*, 7, pp. 85–91.

Sonne, P., 2022. *Ukraine's Zelensky to Russians: "What are you fighting for and with whom?"* [Online] Available at: https://www.washingtonpost.com/national-security/2022/02/23/ukraine-zelensky-russia-address/ [Accessed 23 12 2022].

Sorokowski, A., 1996. Treaty on Friendship, Cooperation, and Partnership between Ukraine and the Russian Federation. *Harvard Ukrainian Studies*, 20, pp. 319–329.

Spetalnick, M., 2008. *Bush to press for Ukraine and Georgia in NATO.* [Online] Available at: https://www.reuters.com/article/uk-nato-ukraine-bush-idUKL0141706220080401 [Accessed 15 01 2023].

Thornton, R., 2015. The Changing Nature of Modern Warfare. *RUSI Journal,* 160(4), pp. 40–48.

Traynor, I., 2008. *Nato allies divided over Ukraine and Georgia.* [Online] Available at: https://www.theguardian.com/world/2008/dec/02/ukraine-georgia [Accessed 24 12 2022].

Traynor, I., 2014. *Ukraine's bloodiest day: Dozens dead as Kiev protesters regain territory from police.* [Online] Available at: https://www.theguardian.com/world/2014/feb/20/ukraine-dead-protesters-police [Accessed 23 03 2023].

U.S. Department of State, 2022. *Russia's war on Ukraine: Six months of lies, implemented.* [Online] Available at: https://www.state.gov/disarming-disinformation/russias-war-on-ukraine-six-months-of-lies-implemented/ [Accessed 17 12 2022].

UN, 2016. *Committee on the Elimination of Racial Discrimination the report of Ukraine.* [Online] Available at: https://www.ohchr.org/en/press-releases/2016/08/committee-elimination-racial-discrimination-report-ukraine [Accessed 25 03 2023].

UN, 2022. *United Nations Charter (Full Text).* [Online] Available at: https://www.un.org/en/about-us/un-charter/full-text [Accessed 19 12 2022].

UNESCO, 2019. *Ukraine 2019 Report.* [Online] Available at: https://en.unesco.org/creativity/node/17027 [Accessed 26 03 2023].

USCC, 2021. *Draft law №5557 on the fundamentals of national resistance presented in Kyiv.* [Online] Available at: https://uscc.org.ua/en/draft-law-5557-on-the-fundamentals-of-national-resistance-presented-in-kyiv/ [Accessed 02 01 2023].

Veidlinger, J., 2022. *Putin's claim to rid Ukraine of Nazis is especially absurd given its history.* [Online] Available at: https://theconversation.com/putins-claim-to-rid-ukraine-of-nazis-is-especially-absurd-given-its-history-177959 [Accessed 23 12 2022].

Verkhovna Rada, 1990. *Declaration of state sovereignty of Ukraine.* [Online] Available at: http://static.rada.gov.ua/site/postanova_eng/ Declaration_of_State_Sovereignty_of_Ukraine_rev1.htm#:~:text= The%20people%20of%20Ukraine%20are, councils%20of%20the%20Ukrainian%20SSR [Accessed 15 01 2023].

Verkhovna Rada, 1991. *Constitution of Ukraine.* [Online] Available at: https://unece.org/fileadmin/DAM/hlm/prgm/cph/experts/ukraine/ ukr.constitution.e.pdf [Accessed 06 01 2023].

Verkhovna Rada, 2015. *On the decision of the National Security and Defense Council of Ukraine dated September 2, 2015 "On the new edition of the Military Doctrine of Ukraine".* [Online] Available at: https://zakon.rada.gov.ua/laws/show/555/2015#Text [Accessed 24 12 2022].

Verkhovna Rada, 2021. *The law of Ukraine on the foundations of national resistance.* [Online] Available at: https://zakon.rada.gov.ua/laws/show/en/1702-20?lang=en#Text [Accessed 01 01 2023].

Vlamis, K., 2022. *Why is Russia attacking Ukraine? Here are 5 reasons Putin and others have given for the invasion.* [Online] Available at: https://www.businessinsider.com/why-russia-is-attacking-ukraine-putin-justification-for-invasion-2022-2?r= US&IR=T [Accessed 26 03 2022].

Von Tunzelmann, A., 2022. *The big idea: Can social media change the course of war?.* [Online] Available at: https://www.theguardian.com/books/2022/apr/25/the-big-idea-can-social-media-change-the-course-of-war [Accessed 20 12 2022].

Walker, E.W., 2016. Between East and West: NATO enlargement and the geopolitics of the Ukraine crisis. In: *Ukraine and Russia. People, Politics, Probaganda and Perspectives.* Bristol: E-International Relations Publising, pp. 134–147.

Ward, A. & Berg, M., 2022. *Give Ukraine long-range missiles: Lithuania's PM.* [Online] Available at: https://www.politico.com/newsletters/national-security-daily/2022/12/06/give-ukraine-long-range-missiles-lithuanias-pm-00072511 [Accessed 15 01 2023].

Warsaw Insitute, 2018. *The Swedish "Total Defence".* [Online] Available at: https://warsawinstitute.org/swedish-total-defence/ [Accessed 23 03 2023].

Weber, J., 2022. *Vladimir Putin's false war claims.* [Online] Available at: https://www.dw.com/en/fact-check-do-vladimir-putins-justifications-for-going-to-war-against-ukraine-add-up/a-60917168 [Accessed 24 12 2022].

White House, The, 2022. *G20 Bali leaders' declaration.* [Online] Available at: https://www.whitehouse.gov/briefing-room/statements-releases/2022/11/16/g20-bali-leaders-declaration/#:~:text=We%20designated%20the%20 G20%20the, time%20of%20unparalleled%20multidimensional%20crises. [Accessed 15 01 2023].

Wilson, A., 2021. *Russia and Ukraine: "One people" as Putin claims?* [Online] Available at: https://rusi.org/explore-our-research/publications/commentary/russia-and-ukraine-one-people-putin-claims [Accessed 17 12 2022].

Wintour, P., 2022a. *Russia strives to avoid G20 isolation as China and India distance themselves.* [Online] Available at: https://www.theguardian.com/world/2022/nov/15/g20-russia-ukraine-war-global-economic-suffering [Accessed 24 12 2022].

Wintour, P., 2022b. *Russia's belief in Nato "betrayal" – and why it matters today.* [Online] Available at: https://www.theguardian.com/world/2022/jan/12/russias-belief-in-nato-betrayal-and-why-it-matters-today [Accessed 21 03 2022].

Yekelchyk, S., 2020. *Ukraine. What Everyone Needs to Know.* 2nd ed. Oxford: Oxford University Press.

Zagoronyuk, A., Frolova, A., Midtunn, H.P., & Pavliuchyk, 2021. *Is Ukraine's reformed military ready to repel a new Russian invasion?.* [Online] Available at: https://www.atlanticcouncil.org/blogs/ukrainealert/is-ukraines-reformed-military-ready-to-repel-a-new-russian-invasion/ [Accessed 02 01 2023].

3 War and Warfare

The Online Environment as a Facilitator of War

Tine Munk

War and Warfare

International Law

When Russian soldiers advanced into Ukraine, other states and citizens were horrified by the atrocities committed by Russia. Recognising that Russia is the aggressor and Ukraine the defender is essential for understanding their positions in international law and customs (Waltzer, 1977, p. 51). Russia chose to attack a sovereign state; it was not forced or left with no other options. The full-scale aggression was a deliberate act that could have been resolved through diplomacy and peaceful negotiations had Russia wished to do so. Under international law, it is illegal for Russia to invade a sovereign state without any justification. The primary international framework for regulating war and warfare is based on customs and treaties developed over the past two centuries, such as the Charter of the UN (1945), the NATO Treaty (1949), the Geneva Conventions (1949), the Geneva Protocol I (1977), were introduced to protect victims in international conflicts. The Hague Conventions (Hague II, 1899 and Hague IV, 1907), these treaties codified existing laws and customs of war and developed new rules to regulate how to conduct war (ICRC, 1899, 1907; Munk, 2022, p. 113; Sterio, 2022).

The Ukrainian defence has obstructed Russia's initial plan for a "lightning strike" attack, forcing troops to advance slowly and inconsistently and to retreat during the Ukrainian counteroffensive. The strong resistance has led to increased Russian attacks targeting civilians, who are otherwise protected by the Hague and Geneva Treaties (Shaw, 2017, p. 898; Solis, 2022, p. 169). The Russian war against Ukraine is not limited to land, air, and sea battles. The battlefield extends to the online environment, computer technologies, networks, and communication platforms. Although conventional warfare is regulated, it becomes problematic when actions occur online in a borderless society without harmonised regulations, prohibition of weapons, or protection of online users and civilians (Munk, 2022, p. 115). The lack of protection from international treaties leaves the area open to abuse by states during war and warfare. Civilian online users are increasingly targeted by online

DOI: 10.4324/9781003432630-3

propaganda and dis-and-misinformation as a war tactic, and measures should be taken to protect them.

Regulating the Online Environment

The Tallinn Manual 1.0 and 2.0 on *International Law Applicable to Cyber Warfare* is a non-binding study developed by an international group of experts under the auspices of NATO to investigate how new forms of information warfare and cyberwarfare are covered by international law. However, these manuals are not part of any doctrines or international agreements (Jones & Kovcich, 2016, p. 299; Munk, 2022, pp. 93, 116). The UN Charter, the Geneva and Hague Conventions, and the NATO Treaty do not address cyber conflicts and attacks as part of ongoing conflicts. While they cover issues such as territorial integrity, armed force, air, land, or sea forces, and armed attacks, there have been no amendments to include cyberspace. Therefore, international law does not regulate online military conflicts such as cyberwarfare, information warfare, or memetic warfare (ICRC, 1899, 1907, 1949, 1977; United Nations, 1945; NATO, 1949; Westby, 2011, p. 86; Munk, 2022, p. 115). The Tallinn Manuals state that the "use of force" can combine cyber and kinetic weapons when their scale and effect are comparable with non-cyber military operations that meet the threshold for using force (Solis, 2022, p. 116).

Russian Justifications

The Russian invasion violates the UN Charter Article 2(4), stating that UN member states should refrain from: "Use of force against the territorial integrity or political independence of any state" (Bellinger III, 2022; UN, 2022).

A central rule of the UN Charter is the prohibition of the threat or use of force, which limits member states' ability to wage war without clear justification (Grey, 2004, p. 29; Janik, 2020, p. 22; Blinken, 2022; De Hoogh, 2022; Sterio, 2022). This provision has only two exceptions: self-defence and actions mandated by the UN Security Council. In this context, the Kremlin relied on two grounds to claim the right to self-defence: one is that Russia received an aid request from the two newly declared independent republics when Russia publicly recognised them on 21 February 2022; the second is that Ukraine threatens Russia's security (Wilmshurst, 2022).

Russia has called on Article 51 of the UN Charter to justify its actions in Ukraine, claiming that it was acting in self-defence against Ukrainian aggression (UN, 2022). However, Russia's actions do not meet the criteria for self-defence under international law, as the Ukrainian military was not threatening Russia, and the invasion was not an immediate response to an armed attack (see Table 3.1) (Grey, 2004, p. 98; Security Council Report, 2022; Wilmshurst, 2022). The argument that a potential enlargement of NATO is a threat to Russia's interests and sovereignty and, therefore, justifies a full-scale invasion is invalid. This is because neither Ukraine nor NATO posed a threat to Russia, and

Table 3.1 The UN Charter Art. 51: The Right to Self-defence

Title	Provision
The Charter of the United Nations 1945. Article 51	Nothing in the present charter shall impair the inherent right of individual or collective self-defence if an armed attack occurs against a member of the United Nations

Source: (Bellinger III, 2022; UN, 2022).

Ukraine was not even on the direct pathway to gaining NATO membership at the time. Additionally, the full-scale Russian invasion did not meet the threshold of imminence for a full-scale invasion (UN, 2022; Wilmshurst, 2022).

Russia's recognition of the two self-declared republics in eastern Ukraine was not recognised by the international community and, therefore, cannot serve as a valid justification (Bellinger III, 2022; UN, 2022). The claims of genocide and discrimination against Russian speakers in Ukraine are unsupported, and Russia's attempt to use Article 51 as a justification for its actions failed. This justification is tied to its claim that Ukraine is committing Russophobia and is associated with fighting neo-Nazis in the Ukrainian government, but there is no evidence to support these claims (Cloud et al., 2022; Li et al., 2022).

Cyberwar and Information Warfare

Hybrid Warfare

Hybrid warfare is a type of warfare that combines both conventional and unconventional means, methods, and combatants, including public and private actors, to blur traditional boundaries between war and warfare, combatants and non-combatants, and offline and online domains (Thornton, 2015, p. 41; Johnson, 2020). This type of warfare employs a full spectrum of capabilities, synchronising means and methods simultaneously to achieve a strategic effect (Bilal, 2021; Munk, 2022, p. 20). Unlike conventional warfare, hybrid warfare involves state and non-state actors operating on multiple levels and blending various attack types and routes to exploit the adversaries' online and offline weaknesses, including doctrinal and strategic vulnerabilities. This type of warfare combines conventional capabilities with irregular tactics and formations, terrorist acts, indiscriminate violence, cohesion, and criminal disorder (Hoffmann, 2009, p. 5; Friedman, 2018; Munk, 2022, p. 22).

In hybrid warfare, various state power tools and instruments are developed during peacetime to be used more coercively during a conflict (Johnson, 2020, p. 228). The actions involved in this strategy often fall below the traditional threshold of war or direct violence. It is a much more accessible, cheaper, and less risky method than conventional military operations. Information warfare and subfields such as propaganda and disinformation play a crucial role in this strategy, adding an extra layer to the offline conflict. Non-military actors, including proxies or civilians, are involved in executing the tactic, making it more challenging to identify and manage due to the lack of a hierarchical structure and outsourcing of planning, execution, and control functions to multiple actors (see Chapters 4 and 5) (Bilal, 2021; Munk, 2022, pp. 116–117).

Cyberwar

Cyberspace and the use of different online platforms are now embedded in military and national cybersecurity strategies. Cyberspace is a global domain that connects and facilitates the functioning of different independent infrastructures by linking Internet telecommunication networks, computer systems, processes, and controllers (Munk, 2022, p. 3). The war has been characterised by Ukraine's use of innovative means and methods to defend against a cyber-equipped Russian cyber force (Munk, 2022; Munk & Ahmad, 2022). The use of cyberspace, online networks, and computer technologies combined with information and information systems have added another layer of complexity to the conflict. The changing nature of war and war-like situations is a product of these technological advances, including the development of cyber weapons, cyborg clones, drones, robots, and IT computerised weapon systems. The innovations and new areas included in modern technology warfare and attacks are often wrongly perceived as a "risk-free" hi-tech "clean war" with "surgical" strikes and minimised casualties. However, the reality is that cyberattacks can cause significant damage, disruption, and loss of life. Integrating cyberspace and computer technologies into warfare strategies requires careful consideration of the risks and potential consequences (Merrin, 2018, p. 42).

In hybrid warfare, known cyber weapons, such as distributed-denial-of-service (DDoS) attacks, defacements, hacking, hack-and-dump, malware, and phishing, are widely used together with online communication forms; use of cyberattacks can have different outcomes spanning from physical warfare to disruption of essential computer systems and spreading spread of dis-and-misinformation, which all provide helpful cyber ammunition when used tactically (Hanna et al., 2021; Munk & Ahmad, 2022). Furthermore, the Internet provides a platform for information warfare on an unprecedented scale. The architecture of the Internet and social media sites enables the widespread use of propaganda and dis-and-misinformation to supplement the technological war through the extended use of numerous platforms hosting and reproducing content (Lavorgna, 2020, p. 172).

Cyberwar offers state actors and governments various opportunities to weaken their opponents. These actors carefully consider their strategic positions between war or warfare situations, where attacks occur frequently using various means and methods but without an impact that would trigger a direct war (Simmons, 2011; Kshetri, 2014, p. 5). Unlike offline warfare, online warfare does not require massive investments in weapons, and states are not bound by treaties and conventions. Economic strength and a passion for weapons often create an unbalanced relationship in the bargaining situation between states. However, these factors are less influential in cyber conflicts because the asymmetric nature of means used in cyberattacks enables actors with limited financial and technical resources to match the powers of more significant and economically developed states (Masters, 2011; Kshetri, 2014, p. 10; Munk, 2022, p. 248). Cyberwarfare is evolving in new directions, such as weaponised memes and communication forms. These less intrusive cyber weapons are produced not to damage the online environment and communication routes in the same way as direct cyberattacks, hacking, malware, etc. Instead, these new ways of exploiting the online environment and interconnectedness are linked to psychological warfare through propaganda and various psychological instruments to influence opinions, emotions, attitudes, and behaviours (see Chapters 4 and 5) (RAND, 2023). The ongoing war against Ukraine illustrates the increasing role of the online environment in the battle. Social media have emerged as a prominent communication pathway, similar to other information warfare means and methods.

Information Warfare

Online information warfare is linked to operations aimed at gaining information advantages over an opponent by controlling the information space and protecting access to one's information. Simultaneously, the goal is to acquire, use, destroy, and disrupt the adversary's information flow during a conflict (NATO, 2023b). The term information warfare refers to:

"[T]he strategic use of information and disinformation to archive political and military goals" (Thornton, 2015; Golovchenko et al., 2018, p. 976).

The online environment has increased and broadened the opportunities to attack and/or reach out to states, organisations, and citizens internally and externally in different areas, such as data acquisition, information defence, and information disruption. With the speed of online communications, it is easy to circulate dis- and-misinformation and other communications with extensive coverage and limited costs (NATO, 2023b).

Information warfare rapidly evolves and has become prominent in defence and policymakers' strategies. Western states, Ukraine, and Russia's conceptualisation of information warfare is linked to waves of weaponised information created and communicated by the state or state-sponsored actors (Golovchenko et al., 2018, p. 977). Various states worldwide take advantage of the new opportunities deriving from the online environment and vital space to get involved in online battles using various computer technologies and software (Molander et al., 1996; Munk, 2022). Cyberwar and information warfare are already central in Russia's war against Ukraine. The Russian forces aggressively use an extensive range of powers to push their agenda by incorporating propaganda, espionage, subversion, and cyberattacks. These methods have been substantially developed and refined since the invasion of the Donetsk and Luhansk regions in 2014 (MacAskill & Johnson, 2016; Kuzio & D'Anieri, 2018, p. 28).

The increased use of social media and virtual spaces is essential in distributing various communications and launching operations controlling the information flow. Social networking platforms are vulnerable to these types of action. For example, trolling is an offensive information warfare tool (see Table 3.2). Many trolling factories employ people to post online comments to promote a specific narrative (NATO, 2023b). The troll farms/factories aim to spread division and create diversions by distributing certain content. Governmental agencies, such as Russia's Internet Research Agency located in Saint Petersburg, constantly distribute propaganda and dis-and-misinformation. The speed and volume of these communications can wear down the most persistent defenders online as the stream of falsehoods is endless, often absurd and unrelated to facts (Seib, 2021, p. 162). During the 2016 US presidential election, Russian troll units, including the Internet Research Agency, produced many memes that circulated on social media discrediting both the US presidential candidate Hillary Clinton and the Democratic Party. These troll factories actively promoted a particular political narrative circulated on social and visual media platforms to influence the outcome (Gault, 2022; Taylor, 2022; Ukrainer, 2022). These memes form a new route within the information warfare framework combined with psychological warfare, aiming to control and progress certain communications, events, or campaigns (Graff, 2018; Thomson & Lapowsky, 2018; Ascott, 2020).

Table 3.2 Key Definitions of Online Methods

Area	Definition
Troll	A person who provokes, disputes, raises controversial topics or attacks online users on social media platforms
Bot	An artificial program that works as an agent for a user or other programs to simulate human activity, such as sending out messages automatically in response to a keyword
Fake news	A weapon by which messages are distributed to mislead online users
War propaganda	Systematic exploitation of the truth, information and their procedural safeguards linked to a persuasion process
Dis-and-misinformation	Disinformation is false information that is deliberately promoted to mislead the audience. Misinformation is false and/or accurate information where the facts are wrongly promoted to an audience

Source: (Scriver, 2015, p. 395; Molina et al., 2021; GOV.UK, 2022; Lutkevich & Gillis, 2022; APS, 2023; NATO, 2023a, 2023b;).

Controlling Online Communication

Information is the first thing to be compromised in war and warfare, where propaganda precedes all sides of the conflict. Some states have developed aggressive strategies to manage the online environment. It is not only authoritarian regimes that attempt to influence online communication and the role of mass media. Democratic governments worldwide also find controlling the internal and external news flow increasingly problematic, particularly the distribution of unintentionally and directly false information (see Table 3.2) (Dunaway & Graber, 2022, p. 19). One of the challenges related to regulating communication is the issue of sovereignty and who sets the rules on modern communication platforms. Sovereignty within international law has a strong territorial dimension that cannot be directly linked to online environments. A nation-state can establish rules for using cyberspace and critical information infrastructure within its jurisdiction (Jones & Kovcich, 2016, p. 301). However, the online environment is borderless, and individuals can move across different jurisdictions without facing border control, much like in the physical world (Liaropoulos, 2012, p. 19). States' territories are defined by geographical, physical, legal, and political constructs, creating a grounded and defined space where sovereignty can be managed exclusively and effectively. States have power in cyberspace by exercising their authority over critical information and communication infrastructure on their territory (Tsagourias, 2021, pp. 14–15). However, the use of social media and virtual spaces is less developed and formalised in an official governmental context. Due to the lack of harmonised legislation and strategies about the territorial scope, responsibilities and powers to control the space, upholding the rule of law and preventing, investigating and prosecuting cybercrimes becomes complicated.

Russian Online War Playbook

The Use of Falsehoods Online

The Russian cyber- and information warfare playbook is known for aggressively pursuing offensive strategies across multiple levels, including the Internet, social media, and virtual spaces. Russian military and intelligence activities promote a disinformation and propaganda ecosystem that has evolved since the 2008 attack on Georgia and has further sharpened and developed since 2014 (Khrebtan-Hörhager & Pyatovskaya, 2022; Paul, 2022). The narratives related to the Russian invasion have been refined, but they still reveal a pattern. For instance, the government in Russia blamed Georgia for ethnic cleansing and distributed Russian passports illegally to protect Russians in South Ossetia (OECD, 2022). The Russian propaganda framework comprises social media operations, overt and covert online proxy media outlets, and state-controlled television and radio programmes that enable the state to promote a specific narrative (U.S. Department of State, 2022). To a certain extent, the model used can be linked to the former Soviet propaganda apparatus, where state-controlled confusion and manipulation were deeply rooted in the tactic. In the Internet era, the Russian model has incorporated current information technology, allowing for the distribution of the narrative through computer technologies and available media (Paul, 2022).

Propaganda and Falsehood

The Russian propaganda machine and influencers are highly active on social media, swiftly responding to any atrocities committed in Ukraine. Their tactic involves dismissing posts as fake or claiming that the Ukrainians have staged events to blame and discredit Russia (Von Tunzelmann, 2022). The Russian information war strategy heavily relies on manipulated photos, false statements, state propaganda and deepfake videos. The battle for the truth is fought on various levels and requires strategic counteraction from both state and civic populations who consume these manipulated visuals and non-visual communications (Wesolowski, 2022). The constant flow of propaganda material and disinformation from the Kremlin has been effective in shaping public opinion in Russia and beyond, creating a reality distortion field that supports the actions of the Russian government.

The propaganda is primarily based on different types of falsehood with distinct features. First, the Russian propaganda machine uses numerous communication channels to distribute messages. Second, Russia will disseminate a high volume of communications based on partial truth or outright fiction to confuse and overwhelm the audience. Third, falsehoods are circulated rapidly and continuously in a repetitive way. Finally, communication lacks consistency (Paul, 2022). Before the annexation of Crimea, Russia intensified its circulation of propaganda and dis-and-misinformation to rewrite history to suit the Kremlin power elite's imperial ambitions. An interesting observation about power and assertiveness is how President Putin is portrayed in memes. Images of Putin often depict him as an action hero, gang leader, and the embodiment of Russia (Denisova, 2020, pp. 89–90).

After the 2014 Russian invasion of the Donetsk and Luhansk regions, Western media painted Russia as the master of propaganda and dis-and-misinformation (Belam, 2018; Ben-David, 2022; Smalley, 2022). Russia employed constant communication flows to pursue a narrative designed to support the war. For example, the Kremlin argued that the annexation of Crimea was not linked to military operations to capture Ukrainian territory. Instead, it promoted a history centred around an uprising by local citizens in Crimea against the Ukrainian government (Golovchenko et al., 2018, p. 976). Russia also followed the well-known narrative of linking Ukraine with Nazism in several memes and other propaganda material. In 2013–2014, the Russian authorities exploited the Nazi theme in mass propaganda to generate a wave of anger towards the Euromaidan protest and used offensive memetic communication to justify the annexation of Crimea and the war in Donetsk and Luhansk (see Chapter 2) (Denisova, 2020, p. 128).

The Russian Playbook Version 2022

In relation to the 2022 invasion, the Kremlin followed the playbook from 2008 and 2014 and distributed a complex narrative on numerous levels. Russian memes were a cornerstone for advancing an aggressive online agenda, which could be linked to Russian military strategies. Before the Russian invasion in February 2022, propaganda and dis-and-misinformation were aggressively circulated to demoralise Ukrainians and create division between the country and its allies. It was also used to enhance public opinion about Russia (OECD, 2022; Wahlstrom et al., 2022). False flags were planted on social media to legitimise the upcoming invasion. For example, online videos showed Russian-backed separatists claiming they needed an immediate evacuation due to unrest in Ukraine (Pereira, 2022). Other information campaigns emerged, arguing that the military buildup by the Ukrainian border was a drill, that Ukraine was an illegitimate state, that neo-Nazis infiltrated the Ukrainian government, that there were severe threats against the Russian minority, and that there was genocide in the Donetsk and Luhansk regions, among other claims. Russia tried to use "whataboutism" to shift the blame from Russia to Ukraine and portray Russia as the victim (OECD, 2022; Wahlstrom et al., 2022).

In the early days of the war, a deepfake video of Zelenskyy was circulated online. The video is a fabrication, but some might believe it (see Table 3.3; Figure 3.1). The video shows President Zelenskyy saying that it was not so easy to be a president and that he would "return to Donbas" after his war effort had failed, and people should lay down arms (BBC News, 2022; Burgess, 2022; Simonite, 2022).

Similarly, pro-Russian online users have tried to discredit the Ukrainian president by spreading online disinformation about his alleged cocaine addiction (see Figure 3.2). Since 24 February 2022, other rumours have emerged, such as claims that Zelenskyy committed suicide or was hospitalised in a critical condition and unable to run the state (Lyngass, 2022; Ukrinform, 2022).

Table 3.3 Text from the Deepfake Video of President Zelenskyy

Speaker	Message
President Zelenskyy (deepfake), 16 March 2022	My advice to you is to lay down arms and return to your families. It is not worth it dying in this war. My advice to you is to live. I am going to do the same

Source: (BBC News, 2022; Burgess, 2022).

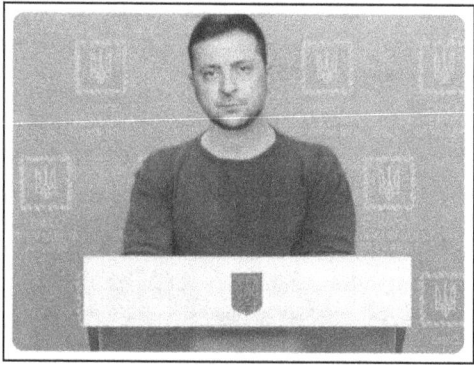

Figure 3.1 Deepfake Video of President Zelenskyy's Supposed Surrender

Pro-Russian accounts have been sharing a digitally altered image of Zelensky appearing to snort cocaine.

The original was taken by an AFP photographer in Bucha, Ukraine in April this year.

Figure 3.2 Fake Claim that President Zelenskyy Has a Cocaine Addiction

Russian trolls also latched onto a bracelet worn by the Commander-in-chief, Zaluzhnyi, in an image of him armed in military camouflage. The trolls claimed to have identified a swastika on the bracelet, which was quickly debunked as fake news. The so-called swastika was a graphic artefact created by enlarging a Celtic eternity knot on the bracelet (see Figure 3.3). However, this meme went viral and fitted into the Kremlin's narrative about fighting neo-Nazis, and the story is still being circulated online (Cymbor, 2022; France24, 2022; Kiennemann, 2022). Another absurd claim that Russian authorities and

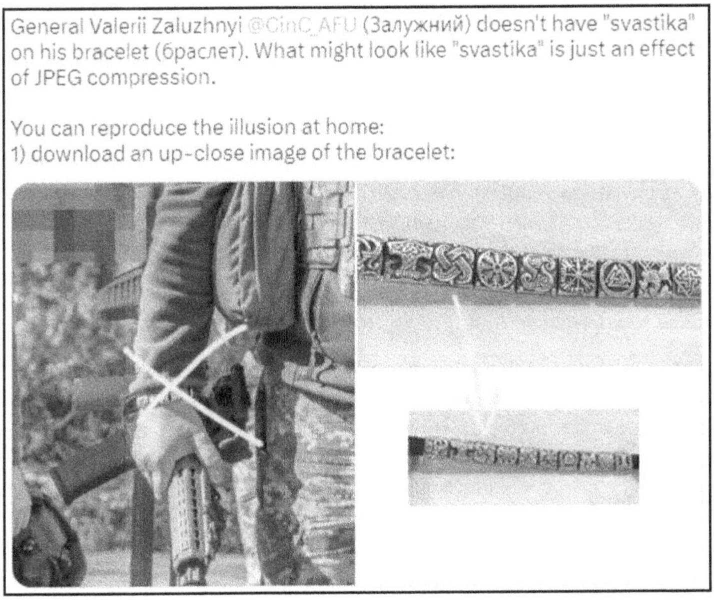

General Valerii Zaluzhnyi @CinC_AFU (Залужний) doesn't have "svastika" on his bracelet (браслет). What might look like "svastika" is just an effect of JPEG compression.

You can reproduce the illusion at home:
1) download an up-close image of the bracelet:

Figure 3.3 Fake Claim that General Zaluzhnyi Wears a Swastika Bracelet

Figure 3.4 Biolab Meme

propaganda repeatedly circulated was the unsupported idea of secret American biolabs placed in Ukraine. The permanent Russian Representative to the UN, Nebenzya, accused the USA and Ukraine of a plot to use migratory birds, bats, and insects to spread pathogens (Borgen et al., 2022). The Russian Federation's military defence spokesperson, Major General Konashenkov, claimed that the USA was training migratory birds to carry Ukrainian bioweapons into Russia (Gilbert, 2022). However, these claims have been used as counter-weapons through various memes ridiculing Russia and even threatening to release the animals from the biolabs (see Figure 3.4).

Access to Information

Authoritarian and democratic states around the world are implementing various strategies to manage the online environment and influence online communication and mass media. However, due to the vast number of online users worldwide and the role of the Internet and social media in their daily lives, subjecting online communication to the same level of regulation and control as offline communication is impossible (European Court of Human Rights, 2022, p. 108). In Russia, the online environment is strictly controlled, with limited access to information on the Internet. The state has been legally able to block websites since the adoption of Russia's Sovereign Internet Law in 2016 (Burgess, 2022). In May 2022, the Kremlin blocked access to Western social media sites, such as Facebook, and other foreign news outlets. The Russian state has also enacted a law criminalising the spread of false information about the Ukrainian invasion, with a penalty of up to 15 years in prison for using words such as "occupation", "invasion", and "war" in reporting about Ukraine (Munk & Ahmad, 2022; Timm, 2022; Troianovski & Safronova, 2022). This law has prevented independent and foreign news outlets from reporting from Russia, with CNN ceasing to broadcast and Bloomberg and the BBC suspending all work inside Russia. *The New York Times* has also removed all reporters from Russia (Tebor, 2022; Timm, 2022).

Information distributed through various media channels, such as television stations, mass media, websites, and social media, etc., reflects a particular political and military reality. In relation to Russian outputs, the quantity and the approximations of the truth by as many different people and outlets as possible are more important than the quality of information (Thornton, 2015, p. 43). Russian forces have targeted Ukraine's mobile communication networks to limit the population's internal and external communication abilities. For example, in March 2022, attacks were launched against TV towers in Kyiv and Kharkiv. Internet connections were also attacked to destabilise the online connections in regions such as Mariupol, Sumy, Kyiv, and Kharkiv (Von Tunzelmann, 2022; Zakir-Hussain, 2022). Targeting communication networks is a tactic authoritarian regimes have used to control information and prevent opposition groups from organising and communicating. The impact of these attacks on Ukraine's communication networks highlights the importance of protecting the Internet and communication infrastructure from attacks, in times of both conflict and peace.

Some of these challenges have been circumvented by using decentralised messaging services and distribution lists and by predominately using social media platforms not controlled by Russia (Musiani, 2022). In February 2022, Elon Musk's Space-X announced on Twitter that Starlink had established an emergency communication link for Ukraine, providing free Internet service to the country's hospitals, schools, and government buildings. This connection was a vital lifeline for Ukraine's people and would help counter Russia's efforts to disrupt communication networks. However, this dependency on Starlink has been problematic, as the company constantly changes the rules for use and disrupts the connection (Fiala, 2022; OECD, 2022; Reuters, 2022; Fingas, 2023; Sabbagh, 2023).

Ukrainian Civic Defence and Online Responses

Ukrainian Legislative Framework

To meet the requirement of the Council of Europe's Convention on Cybercrime 2001 (the Budapest Convention), Ukraine developed a cybersecurity framework to accommodate its responsibilities as a signatory state and develop Ukraine's resilience against attacks. It is clear that Ukraine has taken significant steps to enhance its cybersecurity posture and protect against cyber threats. The Establishment of the National Cybersecurity Coordination Centre (2016) and the adoption of the National Cybersecurity Strategy (2016) are positive developments, as they provide a coordinated approach to cybersecurity and facilitate collaboration between public and private entities (CoE, 2001; Spînu, 2020, p. 4). The Cyber Security Strategy of Ukraine and the Law of Ukraine of Main Basic Principles of Cybersecurity of Ukraine (2017) further support these efforts by providing a legal and organisational foundation for cybersecurity (Verkhovna Rada, 2017).

The amendment of the Cyber Security Strategy in 2020 is a recognition of the evolving threat landscape and the need to adapt to the changing nature of cyberspace. This adaptation is essential to manage the growing information society in Ukraine and safeguard against cyber threats. Similarly, the Doctrine of Information Security (2017) provides a framework to manage disinformation and propaganda, which is critical to combat Russian information aggressions (see Table 3.3) (CCDCOE, 2018; Spînu, 2020, pp. 5–6). The law also sets out the central aims, directions and principles for state and different actors to coordinate the creation of a functional cybersecurity structure. The Doctrine of Information Security creates a framework

Table 3.4 Cybersecurity and the Regulatory Framework

Regulatory framework	Areas covered
The Cyber Security Strategy of Ukraine 2016	Progress the national security system Advance the cyber capabilities actors in the security and defence sector Cybersecurity of critical information and governmental information resources
The Doctrine of Information Security of Ukraine 2017	Define Ukraine's national information interests Identify the threats to the implementation, directions and priorities of state policies regarding the information area Principles of the state information policy formation and implementation; predominately linked to the negative information influence of the Russian Federation Actions related to the conditions of the hybrid war The use of new types of information to influence citizens' consciousness, inciting ethnic and religious hatred, propaganda of aggressive war, offensive change of the constitutional system, or breach of sovereignty/territorial integrity of Ukraine
The Law of Ukraine About the Main Principles of Ensuring Cyber Security of Ukraine 2017	The legal and organisational foundation to protect the key interests of citizens, society, and the state Protect Ukraine's interest in cyberspace Outline the powers and responsibilities of state agencies, enterprises, institutions, organisations, individuals and citizens Set out basic principles of coordination of the actors and agencies' activities, and define basic terms within cybersecurity

Source: (CoE, 2017; GFCE, 2017; Kyiv Post, 2017; Verkhovna Rada, 2018; Spînu, 2020).

for managing Russian information aggressions, propaganda and fake news (CoE, 2017). Overall, the initiatives implemented by Ukraine demonstrate a commitment to enhancing cybersecurity and protecting against cyber threats. However, given the constantly evolving threat landscape, it is essential to continue to adapt and improve cybersecurity measures to stay ahead of potential cyberattacks.

Digital Resilience, Resistance, and Online Management

The concept of resilience of the community (ROC) can indeed be applied to online environments to counter information warfare, especially in the context of national resilience. National resilience requires various measures to protect against threats, including psychological preparedness, identification and reduction of vulnerabilities, and preparation against threats (Fiala, 2022). Building digital resistance is crucial in today's society, where cyber threats are becoming increasingly common. In this regard, Ukraine has taken substantial steps in developing a "whole-of-society" approach where the military and civic actions complement one another. This approach involves protecting the national territory both online and offline by deterring adversaries from invading (see Chapter 2).

With online, hybrid warfare, building resilience in peacetime can be problematic as the security environment constantly changes due to technological developments and threats (Fiala, 2019, p. 196; Munk, 2022, pp. 10, 80). To address this challenge, a proactive and anticipatory approach is necessary, including implementing a foresight system to analyse alternative future directions of technology and communications. A flexible approach to analysing potential threats is essential, which involves monitoring prospective events, warnings and technical developments to create potential threat analysis to guide decision makers about the potential impact of these developments. This approach requires a multi-agency framework that includes public and private actors, replacing strict bureaucratic command-and-control processes with a more flexible approach. Moreover, building resilience in online environments also involves strengthening collaboration and cooperation between stakeholders, including public and private sectors, groups and individuals (Munk, 2015, pp. 84–85).

Resistance in Conflict and War

As a part of ROC, the Ukrainian resistance law requires broad involvement of Ukrainian citizens in irregular warfare (Fiala, 2022). Although there are no explicitly drafted provisions for digital resistance in the Laws on the Foundations of National Resistance 2021 (Verkhovna Rada, 2021) and Law on Amendments to the Law on the Number of the Armed Forces of Ukraine 2021, these provisions can be extended to cover these areas in conjunction with the Cyber Security Strategy (CCDCOE, 2018). President Zelenskyy signed these laws to enhance territorial defence and resistance movements, and they are related to the Special Operations Forces, which are receiving leverage to strengthen defence capabilities in conjunction with other initiatives. The Law on the Fundamentals of National Resistance introduces a system of preparing the population for national resistance and includes an option to activate the Ukrainian population to protect the country, their land and families, and unite citizens around important ideas (President of Ukraine, 2021; USCC, 2021). Article 3(12) is essential because it is directed at countering Russian online information operations (information/memetic warfare) (Verkhovna Rada, 2021). While state actors are involved in digital activism per the Cyber Security Strategy, not all online activities are linked to military or governmental defence operations or are directed by authorities. Private groups and individuals conduct many actions that fall under information warfare (see Chapters 4 and 5).

Online Civic Resistance

Resistance forces utilise various forms of communication, and incorporating Internet communication into the memetic warfare structure provides several advantages, such as speed and accuracy, which aid in early debunking the adversary's communication flow. Civic resistance is a helpful response to counter the propaganda and dis-and-misinformation launched by Russia, and it need not necessarily be within the military framework, but politicians can also participate in such actions. The civil population has a role to play both internally and externally in developing and advancing an active online defence. Civic engagement is essential to raise public awareness about the different issues related to the war. The rise of social media has created a platform for people to voice their opinions and communicate using political and non-political methods that were not traditionally considered part of a country's defence strategy (Smith, 2021, p. 28).

The Total Defence Strategy implemented by Ukraine covers both the civic and military aspects of the country's defence, with civic resistance playing a crucial role (see Chapter 4). Civil resistance refers to the use of coordinated, non-institutional methods and tactics by unarmed individuals to promote change or defence without resorting to violence or aggression towards an opponent (Chenoweth, 2021, p. 2; Justino, 2022). Without public attention and support, Ukraine would have been left powerless in the face of the ongoing conflict (Wolfsfeld, 2022, p. 11). Therefore, the engagement of the population on every level, both internally and externally, is critical to Ukraine's response to the war. The cybersecurity threat extends to the private sector, where civic actors are involved, similar to military units. Online resistance is a combination of both internal and external actors, military and non-military formations, and cooperation. Resilience and resistance are prioritised, leading to comprehensive coordination between civic and military engagement. The use of agile methods enables activities within and beyond government authorisation (Shelest, 2022).

The various facets of cyberwar are not limited to military defence against a specific threat. There is also a need for a societal response that covers all potential dangers. This is where memetic warfare adds a new dimension to defence. The Ukrainian civil society has a clear role embedded in the state's response to Russian aggression, comparable to a multi-agency approach that involves numerous people helping and resisting the hardships of war. This voluntary resistance spans from making food for vulnerable people and army personnel to supporting elderly farmers with their land, animals, and maintenance, as well as other types of support to ensure the country is running smoothly and that those who need help receive it (Onuch & Hale, 2022, pp. 14–15).

There is a tendency for populations and civic movements to come together during times of crisis or war, rallying around the flag or the values of society. However, if the aims and values of the government and the population are not in synergy, this can lead to opposition to government initiatives. Yet, it has been demonstrated that the Russian illegal invasion has brought people closer together and created stronger bonds within communities (see Chapter 5) (Ben-Porath, 2006, p. 25; Pinckney, 2020, p. 21; Onuch & Hale, 2022, p. 251). In Ukraine, politicians and the population have stood together to manage the defence of the country with the support of the population. There has been a change in attitude and perception of how Ukrainians understand their place in the world. They have moved from being a country that needs external security protection to one that is ready to fight for their country with some help from the outside (Kilraine, 2022). This reflects a commitment to the common good and the right to sovereignty, as embodied in civic nationalism. The military doctrine has changed, and the professionalism of the soldiers and the new command line has impacted how the war progresses (see Chapter 2). However, the online battle, including civil memetic warfare, should not be underestimated or brushed off as insignificant. Memetic warfare can manipulate and change opinions and alliances, reaching multiple people simultaneously (NPR, 2018; Fielding & Cobain, 2011; Washington Post, 2022).

Non-violent Actions

The actions taken by civic resistance movements or individuals online are primarily non-violent, using the online environment as the battlefield where memes act as ammunition targeting different groups online. The concept of non-violence is applied outside the usual avenues of politics and without the threat of physical violence (Pinckney, 2021). Non-violence does not mean that actors are passive and submissive; instead, the actions are often defiant, disruptive, and confrontational, depending on the target. The people involved in memetic warfare are a forceful and cohesive group fighting on different levels using tools other than violent or militarised approaches (Chenoweth, 2020; Chenoweth, 2021, p. 35).

Non-violent actions or mobilisation of a large group of people is essential, and participation is often funded by an individual's ethics and civic responsibility (Schock, 2013). The strength of civic engagement in memetic warfare lies in its ability to supplement military actions and undermine Russia's actions and communications. While the civic resistance movement may not hold the same power as the military, its members can reach out to a larger online population to gain support for Ukraine, ridicule Russia's actions, debunk propaganda and dis-and-misinformation, report social media reports, circulating news, and give a boost to morale. The power of memetic warfare is that it can be a preventive measure, rather than just a reactive one, against offensive propaganda and disinformation. In the past, reacting to offensive communications has been a significant challenge in information warfare, and timely responses are crucial to avoid widespread damage from the spread of poisonous information (Seib, 2021, p. 162). The preventive approach involves identifying potential threats and vulnerabilities in advance and taking pro-active steps to address them. This includes developing strategies to counter and disrupt false narratives, promoting alternative narratives, and engaging with target audiences to build resilience. The use of preventive memetic warfare can be a powerful tool in the fight against propaganda and dis-and-misinformation (see Chapter 4).

Memetic Warfare

As a relatively new concept, memetic warfare has become central to the war in Ukraine. On the Russian side, memetic warfare is used as an offensive tactic and is also employed by other groups, such as far-right groups such as the US MAGA movement and terrorist organisations, for example, ISIS. These groups have an online presence and attack different opponents and social groups using memes and social media communication (Mitchell, 2020; Stiegermark, 2020; Donovan et al., 2022). However, memetic warfare as a defensive strategy in war is less defined. Nevertheless, in Ukraine, a pattern of this defensive method has emerged to counter the Russian online presence. The use of memetic warfare includes numerous informal resistance communities, created parallel with military and governmental defence operations.

Memetic warfare includes a concept of warfare mixed with digital and psychological spaces that go beyond what is typically considered cyberwar and information warfare (Erkan, 2022). This concept of memetic warfare falls under the scope of ROC (resilience, operations, and communications) due to its potential to enhance resilience and resistance in preventing, identifying, and responding to asymmetric attacks. These attacks can be an antidote to parts of hybrid warfare where online means and methods destabilise states, polarise society, and spread mistrust (Fiala, 2019, p. 196; Fiala, 2020). Therefore, it is natural that the Ukrainian warfare strategy incorporates similar means and methods as their counterparts to debunk damaging communications or circulate a particular supportive narrative. Memes create a sense of belonging and affiliations where a shared language divides between "them" and "us". The meme culture is also a valuable tool to project a favourable position that can link to a particular social field. Therefore, memes and other visual and non-visual communications can be used to justify judgments, condemnation, and the exclusion of "others" (see Chapters 4 and 5) (Peters & Allan, 2022, p. 218).

The use of humour, parody, and satire in memes can be effective in challenging adversary propaganda. Memes have proved to be effective as a defensive weapon, as they undermine the original purpose of propaganda and disinformation by highlighting their absurdity and providing clarity about the information to a broader audience. Additionally, this creates a new level of resistance by refusing to capitulate to the enemy's false narrative and preventing naive and overly enthusiastic likes without understanding the actual context (Denisova, 2020, pp. 96–97; Chenoweth, 2021, pp. 45–46). Political memes have evolved from occasional glimpses of uncensored personality or concise visual messages to political language used for propaganda, counter-information, or communication with a specific goal in mind (Haddow, 2016). Halting the spread of dis-and-misinformation and preventing the abuse of rhetoric becomes crucial to winning the information war (Denisova, 2020, p. 34; Von Tunzelmann, 2022).

Online communication and memes are strategic means of drawing attention to specific issues and mobilising online users to participate in resistance efforts. This form of communication falls under the umbrella of conventional memetic warfare, which includes political expression and participation in the war without physically engaging in combat. Memetic warfare allows online users to respond in real time to contemporary political events by using the comment fields to progress the communication (Ross & Rivers, 2017, p. 3; Denisova, 2020, p. 34). Additionally, non-visual and visual memes can be considered part of psychological warfare, where various forms of communication are enacted as part of digital propaganda efforts (Ascott, 2020; Gleicher, 2022). In this context, public diplomacy and communications about the war and responses are closely controlled to promote a specific narrative. However, there is also a different side to communication warfare, where memetic warfare can be used in tandem with guerrilla warfare and trolling to control or counter the adversary's information warfare effort (Giesea, 2015, p. 69).

Memes

Memes have become a central part of the Ukrainian online war, transitioning from a conceptual idea to a world of data and pixels. They are easily developed, copied, and changed, incorporating visual and non-visual elements in memes that utilise images, text, and videos (Haddow, 2016). The culture surrounding memes has evolved significantly since Dawkins first coined the term "mimeme" in 1979, which is derived from the ancient Greek word *mimeme* meaning "that which is imitated" or "imitated thing" (Davidson, 2015, p. 121; Miltner, 2017, p. 413; Stiegermark, 2020, p. 110; Barnes et al., 2021, p. 2). Memes are a defined part of internet culture, usually associated with humour and used to gain influence through online transmission. Most memes are created for amusement, expressing personal opinions or mocking something or someone (Miltner, 2014; Miltner, 2017, p. 423). There are various reasons for generating memes, and they can be created either on social media using an image with text or through a meme generator that adds text to an image. Russian-sponsored propaganda and dis-and-misinformation operations have been a part of the geopolitical landscape for over a decade. Their operations have expanded since the Bronze Soldier crisis in Estonia in 2007 and the US presidential election campaign in 2016 (Munk, 2022, pp. 56, 130; Shultz, 2022).

Meme Language

Different types of word are embedded in the memes to identify/define a person, including their language, dialect, slang, organisational jargon, talk choices, and accents. These written and spoken languages can act as identifiers for group membership and can be either a part of the in-group or the out-group depending on these markers (see Chapter 4) (Haji et al., 2016, p. xv; Keblusek et al., 2017, p. 2). As memes have taken on the form of hashtags, tweets, photos, quotes, or jokes that are repetitively shared online, they are highly visible to audiences on various platforms. Memes are transmitted from person to person through different media sites, such as social media, online news, videos, or blog posts. Once uploaded, they can reach a vast global audience within a short time span. These viral memes are essential shared social phenomena that represent common opinions, cultural norms, and political power, or are used to advance societal change (Dynel & Messerli, 2020; Barnes et al., 2021, p. 2).

Direct communications (non-visual/ visual) are essential means to keep people informed and enhance group cohesion. For instance, visual communication, such as photographs and videos containing important governmental speeches, are distributed to large groups online using platforms such as Twitter, Instagram, and TikTok. Memes are also used as communication campaign tactics, cultivating para-social relationships between political figures and their followers. Through different forms of social media, the audience is encouraged to actively participate by commenting and sharing their memes and content across various platforms (Lipschultz, 2022, p. 94).

Social Media Posts

Ukrainian social media posts play a significant role in countering the Russian disinformation campaign. Ukrainian citizens are skilled communicators who document and share their victories, live stream events, and capture various actions, making it challenging to manipulate people with disinformation (Pereira, 2022). In 2022, social media and instant messaging had an even more prominent position in everyday life compared to 2014, when the war began. Online use has given people unprecedented insight into the conflict, with raw images and videos of troop movements, equipment transport, civilian casualties, and destruction of Ukrainian cities broadcast more or less live. The Russian invasion and Ukrainian response are well-documented online, with a 24/7 live display of the war, which is unprecedented compared to previous conflicts (Lopatto, 2022; Shultz, 2022).

People who use social media are often tech savvy and have extensive online networks. Posts are constantly being uploaded from both in-groups and out-groups, and essential communications can be routed through these social networks. Often, a post can provide valuable information to authorities or military forces about advisory movements or locations. Gathering information and intelligence from different online sources and communicating on the platforms are also parameters included in memetic warfare. Public and private information can be helpful for intelligence services and military units (Bernot & Childs, 2022). Oversharing information regarding the locations of soldiers and equipment online can be fatal, especially with geotagged social media posts. Therefore, this information should be communicated to authorities, and caution should be exercised when sharing sensitive information online (Yeung & Oliker, 2015; Suciu, 2022).

The strategic use of memes extends beyond everyday vocabulary, as they are embedded as symbolic rhetorical arguments in various conversations and debates. For example, in January 2023, the #FreeTheLeopards campaign went viral, supporting arguments to pressure and embarrass Germany to allow countries to support the Leopard 2 battle tanks in Ukraine. Social media groups and individuals activated an intensive online campaign using their networks to pressure Germany (see Figures 3.5, 3.6). This successful campaign

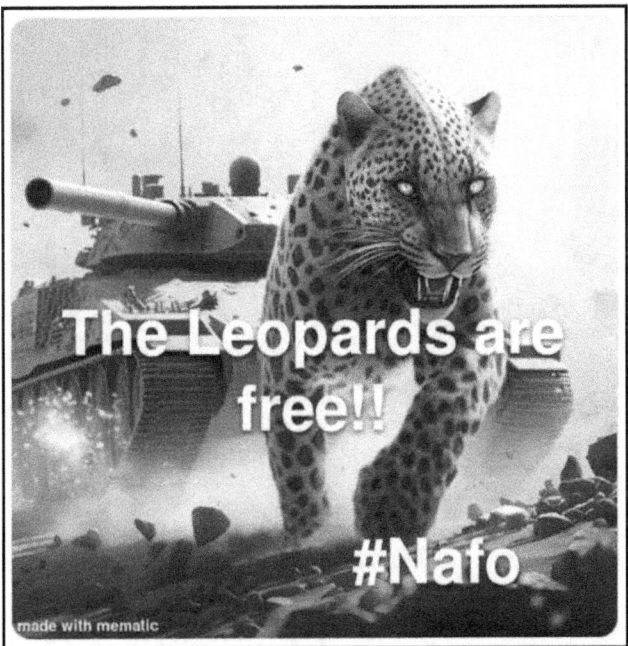

Figure 3.5 Meme Celebrating the Leopards

Figure 3.6 Meme Imagining the Free Leopards

was partly conducted through political pressure from Ukraine's political allies, such as Finland, Sweden, Slovakia, and Denmark. The stalemate about
heavy weapons and tanks ended when German Chancellor Scholz confirmed
that countries could send tanks to Ukraine (Demianyk, 2023; Mac Dougall,
2023; Smith, 2023).

References

APS, 2023. *Misinformation and disinformation.* [Online] Available at: https://www. apa.org/ topics/journalism-facts/misinformation-disinformation#:~:text= Misinformation%20is%20false%20or%20inaccurate, intentionally% 20making%20the%20 misstating%20facts. [Accessed 19 03 2023].

Ascott, T., 2020. *How memes are becoming the new frontier of information warfare.* [Online] Available at: https://www.aspistrategist.org.au/how-memes-are-becoming-the-new-frontier-of-information-warfare/ [Accessed 25 12 2022].

Barnes, K., Riesenmy, T., Trinh, M.D. & Lleshi, E., 2021. Dank or Not? Analyzing and predicting. *Applied Network Science*, 6(1).

BBC News, 2022. *Deepfake presidents used in Russia-Ukraine war.* [Online] Available at: https://www.bbc.co.uk/news/technology-60780142 [Accessed 18 03 2023].

Belam, M., 2018. *Twitter diplomacy: How Russian embassy trolls UK government.* [Online] Available at: https://www.theguardian.com/uk-news/2018/mar/15/twitter-diplomacy-how-russian-embassy-trolls-british-government [Accessed 27 12 2022].

Bellinger III, J.B., 2022. *How Russia's invasion of Ukraine violates international law.* [Online] Available at: https://www.cfr.org/article/how-russias-invasion-ukraine-violates-international-law#:~:text=An%20Illegal%20Invasion, political% 20independence%20of%20any%20state.%E2%80%9D [Accessed 19 12 2022].

Ben-David, D., 2022. *Russia's UK embassy tweets antisemitic meme of Ukraine's President Zelensky.* [Online] Available at: https://www.thejc.com/news/world/ russia's-uk-embassy-tweets-antisemitic-meme-of-ukraines-president-zelensky-1nhVLHqg6ripjCNfyKWZsl [Accessed 27 12 2022].

Ben-Porath, S.R., 2006. *Citizenship Under Fire: Democratic Education in Times of Conflict.* Princeton: Princeton University Press.

Bernot, A. & Childs, A., 2022. *Social media in times of war.* [Online] Available at: https://www.lowyinstitute.org/the-interpreter/social-media-times-war [Accessed 16 03 2023].

Bilal, A., 2021. *Hybrid warfare – new threats, complexity, and "trust" as the antidote.* [Online] Available at: https://www.nato.int/docu/review/articles/ 2021/11/30/hybrid-warfare-new-threats-complexity-and-trust-as-the-antidote/index.html [Accessed 19 01 2023].

Blinken, A.J., 2022. *The stakes of Russian aggression for Ukraine and beyond.* [Online] Available at: https://www.state.gov/the-stakes-of-russian-aggression-for-ukraine-and-beyond/ [Accessed 19 12 2022].

Burgess, M., 2022. *Russia is quietly ramping up its internet censorship machine.* [Online] Available at: https://www.wired.co.uk/article/russia-internet-censorship-splinternet [Accessed 06 01 2023].

Burgess, S., 2022. *Ukraine war: Deepfake video of Zelenskyy telling Ukrainians to "lay down arms" debunked.* [Online] Available at: https://news.sky.com/story/ ukraine-war-deepfake-video-of-zelenskyy-telling-ukrainians-to-lay-down-arms-debunked-12567789 [Accessed 07 01 2023].

CCDCOE, 2018. *Cyber Security Strategy of Ukraine.* [Online] Available at: https://ccd-coe.org/uploads/2018/10/NationalCyberSecurityStrategy_Ukraine.pdf [Accessed 04 01 2022].

Chenoweth, E., 2020. The Future of Nonviolent Resistance. *Journal of Democracy*, 31(3), pp. 69–84.

Chenoweth, E., 2021. *Civil Resistance. What Everyone Needs to Know.* Oxford: Oxford University Press.

Cloud, D., Marson, J., Gershkovich, E., & Norman, L., 2022. *Russia ramps up accusations of neo-Nazism in Ukraine.* [Online] Available at: https://www.wsj.com/articles/russia-reiterates-accusations-of-neo-nazism-in-ukraine-11651579622 [Accessed 15 03 2023].

CoE, 2001. *Convention on Cybercrime.* [Online] Available at: https://rm.coe.int/1680081561 [Accessed 27 03 2022].

CoE, 2017. *Doctrine of Information Security Ukraine.* [Online] Available at: https://rm.coe.int/doctrine-of-information-security-of-ukraine-developments-in-membersta/168073e052 [Accessed 18 01 2023].

Cymbor, P., 2022. *No, Valerii Zaluzhnyi, Commander-in-Chief of the Armed Forces of Ukraine, does not wear a bracelet with a swastika.* [Online] Available at: https://fakenews.pl/en/general/no-valerii-zaluzhnyi-commander-in-chief-of-the-armed-forces-of-ukraine-does-not-wear-a-bracelet-with-a-swastika/ [Accessed 10 10 2023].

Davidson, P., 2015. The Social Media Reader. In: *The Social Media Reader.* s.l.:s.n., pp. 120–134.

De Hoogh, A., 2022. *The elephant in the room: Invoking and exercising the right of collective self-defence in support of Ukraine against Russian aggression.* [Online] Available at: https://opiniojuris.org/2022/03/07/the-elephant-in-the-room-invoking-and-exercising-the-right-of-collective-self-defence-in-support-of-ukraine-against-russian-aggression/ [Accessed 19 12 2022].

Demianyk, G., 2023. *Free the Leopards: Why German tanks are so prized by Ukraine.* [Online] Available at: https://www.huffingtonpost.co.uk/entry/leopard-tanks-germany-us-ukraine_uk_63d05c05e4b0c8e3fc7a6d0c [Accessed 28 01 2023].

Denisova, A., 2020. *Internet Memes and Society. Social, Cultural and Political Context.* London: Routledge.

Donovan, J., Dreyfuss, E., & Friedberg, B., 2022. *Meme Wars.* New York: Bloomsbury Publishing.

Dunaway, J.L. & Graber, D.A., 2022. *Mass Media and American Politics.* 11th ed. s.l. London: Sage.

Dynel, M. & Messerli, T.C., 2020. On a Cross-cultural Memescape: Switzerland through nation memes from within and from the outside. *Contrastive Pragmat,* 1, pp. 210–241.

European Court of Human Rights, 2022. *Guide on Article 10 of the European Convention on Human Rights. Freedom of Expression,* Stasbourg: CoE.

Fiala, O., 2019. *Resitance Operating Concept.* Stockholm: Special Operations Command Europe (SOCEUR) & the Swedish Defence University.

Fiala, O.C., 2020. *ROC. Resistance Operating Concept.* MacDill Air Force Base(Florida): JSOU Press.

Fiala, O., 2022. Resilience and Resistance in Ukraine. *Small Wars Journal,* 31, p. 12.

Fielding, N. & Cobain, I., 2011. *Revealed: US spy operation that manipulates social media.* [Online] Available at: https://www.theguardian.com/technology/2011/mar/17/us-spy-operation-social-networks [Accessed 05 01 2023].

Fingas, J., 2023. *SpaceX doesn't want Ukraine using Starlink to control military drones.* [Online] Available at: https://www.engadget.com/spacex-objects-to-ukraine-starlink-drones-215629197.html?guccounter=1&guce_referrer=aHR0cHM6Ly93d3cuZ29v Z2xlLmNvbS88&guce_referrer_sig=AQAAAMqxSE4OHDSmPhmMnGfG_

yi5swkQYOSLgfVkrCFk_fKwhuh5e8Qc6BzakW9apn8fEqwfRoKBk84-pquooJprhcLv-fTF [Accessed 29 03 2023].

France24, 2022. *The Ukrainian commander-in-chief's "swastika" bracelet.* [Online] Available at: https://www.france24.com/en/tv-shows/truth-or-fake/20221012-we-take-a-look-at-the-ukrainian-commander-in-chief-s-swastika-bracelet [Accessed 07 01 2023].

Friedman, O., 2018. *Russian Hybrid Warfare.* Oxford: Oxford University Press.

Gault, M., 2022. *Shitposting Shiba Inu accounts chased a Russian diplomat offline.* [Online] Available at: https://www.vice.com/en/article/y3pd5y/shitposting-shiba-inu-accounts-chased-a-russian-diplomat-offline [Accessed 27 12 2022].

GFCE, 2017. *Cybersecurity in Ukraine: National strategy and international cooperation.* [Online] Available at: https://thegfce.org/cybersecurity-in-ukraine-national-strategy-and-international-cooperation/ [Accessed 18 01 2023].

Giesea, J., 2015. It's Time to Embrace Memetic Warfare. *Defence Strategic Communications*, 1(1), pp. 67–75.

Gilbert, D., 2022. *Russia is now claiming the US trained birds to deliver Ukrainian bioweapons.* [Online] Available at: https://www.vice.com/en/article/7kbmgg/russia-ukraine-bioweapons-trained-birds [Accessed 30 11 2022].

Golovchenko, Y., Hartmann, M., & Adler-Nissen, R., 2018. State, Media and Civil Society in the Information Warfare over Ukraine: Citizen curators of digital disinformation. *Yevgeniy Golovchenko, Mareike Hartmann, Rebecca Adler-Nissen, International Affairs*, 94(5), pp. 975–994.

GOV.UK, 2022. *UK exposes sick Russian troll factory plaguing social media with Kremlin propaganda.* [Online] Available at: https://www.gov.uk/government/news/uk-exposes-sick-russian-troll-factory-plaguing-social-media-with-kremlin-propaganda [Accessed 04 01 2023].

Graff, G. M., 2018. *Russian trolls are still playing both sides – even with the Mueller probe.* [Online] Available at: https://www.wired.com/story/russia-indictment-twitter-facebook-play-both-sides/ [Accessed 25 12 2022].

Grey, C., 2004. *International Law and the Use of Force.* 2nd ed. Oxford: Oxford University Press.

Haddow, D., 2016. *Meme warfare: how the power of mass replication has poisoned the US election.* [Online] Available at: https://www.theguardian.com/us-news/2016/nov/04/ political-memes-2016-election-hillary-clinton-donald-trump [Accessed 25 2 2022].

Haji, R., McKeown, S., & Ferguson, N., 2016. Social Identity and Peace Psychology. In: *Understanding Peace and Conflict Through Social Identity Theory.* Cham: Springer Nature, pp. XV–XX.

Hanna, K.T., Ferguson, K., & Rosencrance, l., 2021. *Cyberwarfare.* [Online] Available at: https://www.techtarget.com/searchsecurity/definition/cyberwarfare [Accessed 18 09 2022].

Hoffmann, F.G., 2009. Hybrid Threats: Reconceptualisaing the evolving character of modern conflicts. *Strategic Forum*, 4, pp. 1 –8.

ICRC, 1899. *Convention (II) with Respect to the Laws and Customs of War on Land and its Annex: Regulations concerning the Laws and Customs of War on Land.* [Online] Available at: https://ihl-databases.icrc.org/ihl/INTRO/150 [Accessed 03 04 2022].

ICRC, 1907. *Convention (IV) Respecting the Laws and Customs of War on Land and its annex: Regulations concerning the Laws and Customs of War on Land. The Hague,*

18 October 1907. [Online] Available at: https://ihl-databases.icrc.org/en/ihl-treaties/ hague-conv-iv-1907?activeTab=undefined [Accessed 19 12 2022].

ICRC, 1949. *The Geneva Conventions of 12 August 1949.* [Online] Available at: https:// www.icrc.org/en/doc/assets/files/publications/icrc-002-0173.pdf [Accessed 29 12 2020].

ICRC, 1977. *Protocol Additional to the Geneva Conventions of 12 August 1949, and Relating to the Protection of Victims of International Armed Conflicts (Protocol I), 8 June 1977.* [Online] Available at: https://ihl-databases.icrc.org/ihl/INTRO/470 [Accessed 29 12 2020].

Janik, R., 2020. *International Law and the Use of Force.* 1st ed. London: Routledge.

Johnson, R., 2020. Military Strategy for Hybrid Confrontation and Coercion. In: *Military Strategy in the Twenty-First Century. The Challenge for NATO.* London: Hurst, pp. 227–249.

Jones, A. & Kovcich, G., 2016. *Global Information Warfare: The new digital battlefield.* 2nd ed. Boca Raton: CRC Press.

Justino, P., 2022. *The war in Ukraine: Civilian vulnerability, resilience, and resistance.* [Online] Available at: https://cepr.org/voxeu/columns/war-ukraine-civilian-vulnerability-resilience-and-resistance [Accessed 29 01 2023].

Keblusek, L., Giles, H., & Maass, A., 2017. Communication and Group Life: How language and symbols shape intergroup relations. *Group Processes and Intergroup Relations,* 20(5), pp. 1–12.

Khrebtan-Hörhager, J. & Pyatovskaya, E., 2022. *Putin's propaganda is rooted in Russian history – and that's why it works.* [Online] Available at: https://theconversation.com/ putins-propaganda-is-rooted-in-russian-history-and-thats-why-it-works-184197 [Accessed 04 01 2023].

Kiennemann, L., 2022. *No, this Ukrainian general isn't wearing a bracelet with a swastika on it.* [Online] Available at: https://uk.finance.yahoo.com/news/no-ukrainian-general-isn-t-155713103.html?guccounter=1&guce_referrer=aHR0cHM6Ly93d3cuZ 29vZ2xlLmNvbS8&guce_referrer_sig=AQAAAFSXsNkqarndlmLhXBhDz-SYyL5Hf0Wx8fQaDohe_DqK2dlJpNuLyRc6ySK49rD8LYUw5K_pSG35lSisU-UAsOG4d7D5EjU229 [Accessed 07 01 2023].

Kilraine, L., 2022. *Humour is crucial weapon in Ukraine's online war, says media specialist.* [Online] Available at: https://www.belfasttelegraph.co.uk/news/uk/ humour-is-crucial-weapon-in-ukraines-online-war-says-media-specialist-42138670.html [Accessed 28 12 2022].

Kshetri, N., 2014. *Cybersecurity and International Relations: The U.S. engagement with China and Russia.* Greensboro: s.n.

Kyiv Post, 2017. *Poroshenko signs law on key principles of ensuring Ukraine's cyber security.* [Online] Available at: https://www.kyivpost.com/ukraine-politics/ poroshenko-signs-law-key-principles-ensuring-ukraines-cyber-security.html [Accessed 18 01 2023].

Li, D., Allen, J., & Siemaszko, C., 2022. *Putin using false "Nazi" narrative to justify Russia's attack on Ukraine, experts say.* [Online] Available at: https://www.nbcnews. com/news/world/putin-claims-denazification-justify-russias-attack-ukraine-experts-say-rcna17537 [Accessed 15 03 2023].

Liaropoulos, A., 2012. Exercising State Sovereignty in Cyberspace: An international cyber-order under construction. *Journal of Information Warfare,* 12(2), pp. 19–26.

Lipschultz, J.H., 2022. *Social Media and Political Communication.* London: Taylor & Francis.

Lopatto, E., 2022. *The limits of Putin's propaganda.* [Online] Available at: https://www.wilsoncenter.org/blog-post/limits-putins-propaganda [Accessed 31 12 2022].

Lutkevich, L. & Gillis, A.S., 2022. *Bot.* [Online] Available at: https://www.techtarget.com/whatis/definition/bot-robot [Accessed 04 01 2023].

Lyngass, S., 2022. *Pro-Russia online operatives falsely claimed Zelensky committed suicide in an effort to sway public opinion, cybersecurity firm says.* [Online] Available at: https://edition.cnn.com/2022/05/19/politics/pro-russia-disinformation-report/index.html [Accessed 07 01 2023].

Mac Dougall, D., 2023. *"Free the Leopards!" Campaign aims to "embarrass" Germany into sending tanks to Ukraine.* [Online] Available at: https://www.euronews.com/2023/01/05/free-the-leopards-campaign-aims-to-embarrass-germany-into-sending-tanks-to-ukraine [Accessed 28 01 2023].

Masters, J., 2011. *Confronting the cyber threat.* [Online] Available at: https://www.cfr.org/backgrounder/confronting-cyber-threat [Accessed 16 11 2020].

Merrin, W., 2018. *Digital War.* Abingdon: Routledge.

Miltner, K.M., 2014. "There's No Place for Lulz on LOLCats": The role of genre, gender, and group identity in the interpretation and enjoyment of an Internet meme. *First Monday*, 19(8).

Miltner, K.M., 2017. Internet Memes. In: *The Sage Handbook of Social Media.* London: Sage.

Mitchell, G., 2020. *Memetic Warfare.* s.l.: Running-Bull Publishing.

Molander, R.C., Riddile, A., & Wilson, P. A., 1996. *Strategic information warfare. A new face of war.* [Online] Available at: https://www.rand.org/pubs/monograph_reports/MR661.html [Accessed 17 09 2022].

Molina, M., Sundar, S.S., Le, T., & Lee, D., 2021. "Fake News" Is Not Simply False Information: A concept explication and taxonomy of online content. *American Behavioral Scientist*, 65(2), pp. 180–212.

Munk, T., 2022. *The Rise of Politically Motivated Cyber Attacks.* London: Routledge.

Munk, T. & Ahmad, J., 2022. "I Need Ammunition, Not a Ride": The Ukrainian cyber war. *Comunicação e Sociedade*, 42, pp 221–241.

Musiani, F., 2022. *Resistance in Ukraine is also digital.* [Online] Available at: https://news.cnrs.fr/opinions/resistance-in-ukraine-is-also-digital [Accessed 13 11 2022].

NATO, 1949. *The North Atlantic Treaty.* [Online] Available at: https://www.nato.int/cps/ en/natolive/official_texts_17120.htm [Accessed 16 12 2020].

NATO, 2023a. *Media -(dis)information - security.* [Online] Available at: https://www.nato.int/nato_static_fl2014/assets/pdf/2020/5/pdf/2005-deepportal2-troll-factories.pdf [Accessed 04 01 2023].

NATO, 2023b. *Media - (dis)information - security.* [Online] Available at: https://www.nato.int/nato_static_fl2014/assets/pdf/2020/5/pdf/2005-deepportal4-information-warfare.pdf [Accessed 03 01 2023].

NPR, 2018. *The "weaponization" of social media — and its real-world consequences.* [Online] Available at: https://www.npr.org/2018/10/09/655824435/ the-weaponization-of-social-media-and-its-real-world-consequences [Accessed 05 01 2023].

OECD, 2022. *Disinformation and Russia's war of aggression against Ukraine.* [Online] Available at: https://www.oecd.org/ukraine-hub/policy-responses/disinformation-and-russia-s-war-of-aggression-against-ukraine-37186bde/ [Accessed 03 01 2023].

Onuch, O. & Hale, H.E., 2022. *The Zelensky Effect.* London: Hurst.

Paul, K., 2022. *Flood of Russian misinformation puts tech companies in the hot seat.* [Online] Available at: https://www.theguardian.com/media/2022/feb/28/facebook-twitter-ukraine-russia-misinformation [Accessed 20 12 2022].

Pereira, I., 2022. *Memes become weapons in Ukraine-Russia conflict.* [Online] Available at: https://abcnews.go.com/International/memes-weapons-ukraine-russia-conflict/story?id=83184578 [Accessed 19 09 2022].

Peters, C. & Allan, S., 2022. Weaponizing Memes: The journalistic mediation of visual politicization. *Digital Journalism*, 10(2), pp. 217–229.

Pinckney, J.C., 2020. *From Dissent to Democracy.* Oxford: Oxford University Press.

Pinckney, J., 2021. Nonviolent Resistance, Social Justice, and Positive Peace. In: *The Palgrave Handbook of Positive Peace.* London: Palgrave Macmillan, pp. 1–16.

President of Ukraine, 2021. *President signed laws on national resistance and increasing the number of the Armed Forces.* [Online] Available at: https://www.president.gov.ua/en/news/glava-derzhavi-pidpisav-zakoni-pro-nacionalnij-sprotiv-i-zbi-69809 [Accessed 04 01 2023].

RAND, 2023. *Psychological warfare.* [Online] Available at: https://www.rand.org/topics/psychological-warfare.html [Accessed 20 01 2023].

Reuters, 2022. *Starlink helped restore energy, communications infrastructure in parts of Ukraine – official.* [Online] Available at: https://www.reuters.com/world/starlink-helped-restore-energy-communications-infrastructure-parts-ukraine-2022-10-12/ [Accessed 13 11 2022].

Ross, A.S. & Rivers, D.J., 2017. Digital Cultures of Political Participation: Internet memes and thediscursive delegitimization of the 2016 U.S presidential candidates. *Discourse, Context & Media*, 16, pp. 1–11 (https://www.sciencedirect.com/science/article/abs/pii/S2211695816301684? via%3Dihub)

Sabbagh, D., 2023. *Fury in Ukraine as Elon Musk's SpaceX limits Starlink use for drones.* [Online] Available at: https://www.theguardian.com/world/2023/feb/09/zelenskiy-aide-takes-aim-at-curbs-on-ukraine-use-of-starlink-to-pilot-drones-elon-musk [Accessed 23 02 2023].

Schock, K., 2013. The Practice and Study of Civil Resistance. *Journal of Peace Research*, 50(3), pp. 277–290.

Scriver, S., 2015. War Probaganda. In: *International Encyclopedia of the Social & Behavioral Sciences.* 2nd ed. Oxford: Elsevier, pp. 395–400.

Security Council Report, 2022. *In hindsight: Ukraine and the tools of the UN.* [Online] Available at: https://www.securitycouncilreport.org/monthly-forecast/2022-03/in-hindsight-ukraine-and-the-tools-of-the-un.php [Accessed 06 01 2023].

Seib, P., 2021. *Information at War.* 1st ed. Cambridge: Polity Press.

Shaw, M., 2017. *International Law.* 8th ed. Cambridge: Cambridge Univesity Press.

Shelest, H., 2022. *Defend. Resist. Repeat: Ukraine's lessons for European defence.* [Online] Available at: https://ecfr.eu/publication/defend-resist-repeat-ukraines-lessons-for-european-defence/ [Accessed 26 12 2022].

Shultz, B., 2022. *Memew: What western governments can learn from the NAFO alliance.* [Online] Available at: https://intpolicydigest.org/meme-warfare-what-western-governments-can-learn-from-the-nafo-alliance/ [Accessed 31 12 2022].

Simmons, B. A., 2011. International Studies in the Global Information Age. *International Studies Quarterly*, 55(3), pp. 589–599.

Simonite, T., 2022. *A Zelensky deepfake was quickly defeated. The next one might not be.* [Online] Available at: https://www.wired.com/story/zelensky-deepfake-facebook-twitter-playbook/ [Accessed 07 01 2023].

Smalley, S., 2022. *How one group of "fellas" is winning the meme war in support of Ukraine.* [Online] Available at: https://www.cyberscoop.com/nafo-fellas-and-their-memes-ukraine/ [Accessed 27 12 2022].

Smith, P., 2023. *Ukraine to get Leopard tanks from Germany, ending a rift among Western allies.* [Online] Available at: https://www.nbcnews.com/news/world/ukraine-leopard-tanks-germany-rift-allies-russia-war-rcna66943 [Accessed 28 01 2023].

Smith, R.L., 2021. *Civic Engagement. Making a Change Together.* s.l.: Creative Book Writers.

Solis, G.D., 2022. *The Law of Armed Conflict. International Humanitarian Law in War.* Cambridge: Cambridge University Press.

Spînu, N., 2020. *Ukraine Cybersecurity Governance Assessment.* Geneva: DCAF.

Sterio, M., 2022. *The Russian invasion of Ukraine: Violations of international law.* [Online] Available at: https://www.jurist.org/commentary/2022/07/milena-sterio-russia-war-crimes-ukraine/ [Accessed 19 12 2022].

Stiegermark, A., 2020. *Memetic Warfare.* Helsingborg: Logik.

Suciu, P., 2022. *Military personnel need watch how they use social media.* [Online] Available at: https://www.forbes.com/sites/petersuciu/2022/07/15/military-personnel-need-watch-how-they-use-social-media/?sh=697051525943 [Accessed 04 01 2023].

Taylor, A., 2022. *With NAFO, Ukraine turns the trolls on Russia.* [Online] Available at: https://www.washingtonpost.com/world/2022/09/01/nafo-ukraine-russia/ [Accessed 01 09 2022].

Tebor, C., 2022. *Russia increases censorship with new law: 15 years in jail for calling Ukraine invasion a "war".* [Online] Available at: https://eu.usatoday.com/story/news/world/ukraine/2022/03/08/russia-free-speech-press-criminalization-misinformation/9433112002/ [Accessed 05 01 2023].

Thomson, N. & Lapowsky, I., 2018. *How Russian trolls used meme warfare to divide America.* [Online] Available at: https://www.wired.com/story/russia-ira-propaganda-senate-report/?redirectURL=%2Fstory%2Frussia-ira-propaganda-senate-report%2F [Accessed 25 12 2022].

Thornton, R., 2015. The Changing Nature of Modern Warfare. *The RUSI Journal,* 160(4), pp. 40–48.

Timm, T., 2022. *In its quest to censor war reporting, the Russian government has dismantled all semblance of press freedom.* [Online] Available at: https://freedom.press/news/in-its-quest-to-censor-war-reporting-the-russian-government-has-dismantled-all-semblance-of-press-freedom/ [Accessed 19 09 2022].

Troianovski, A. & Safronova, V., 2022. *Russia takes censorship to new extremes, stifling war coverage.* [Online] Available at: https://www.nytimes.com/2022/03/04/world/europe/russia-censorship-media-crackdown.html [Accessed 19 09 2022].

Tsagourias, N., 2021. The Legal Status of Cyberspace: Sovereignty redux?. In: *Research Handbook on International Law and Cyberspace.* Cheltenham: Edward Elgar Publishing, pp. 9–31.

U.S. Department of State, 2022. *Russia's top five persistent disinformation narratives.* [Online] Available at: https://www.state.gov/russias-top-five-persistent-disinformation-narratives/ [Accessed 04 01 2023].

Ukrainer, 2022. *Who are the NAFO Fellas? The army of cartoon dogs fighting Russian propaganda.* [Online] Available at: https://ukrainer.net/nafo-fellas/ [Accessed 27 12 2022].

Ukrinform, 2022. *President Zelensky debunks fake news about his health condition.* [Online] Available at: https://www.ukrinform.net/rubric-polytics/3534016-president-zelensky-debunks-fake-news-about-his-health-condition.html [Accessed 07 01 2023].

United Nations, 1945. *Charter of the United Nations 1945.* [Online] Available at: https://www.un.org/en/sections/un-charter/un-charter-full-text/ [Accessed 07 12 2020].

United Nations, 2022. *United Nations Charter (full text).* [Online] Available at: https://www.un.org/en/about-us/un-charter/full-text [Accessed 19 12 2022].

USCC, 2021. *Ukrainian parliament (Verkhovna Rada) registered a bill #5557 "About foundations of national resistance".* [Online] Available at: https://uscc.org.ua/en/ukrainian-parliament-verkhovna-rada-registered-a-bill-5557-about-foundations-of-national-resistance/ [Accessed 04 01 2023].

Verkhovna Rada, 2017. *Law of Ukraine on the Basic Principles of Cybersecurity in Ukraine.* [Online] Available at: https://zakon.rada.gov.ua/laws/show/en/2163-19#Text [Accessed 03 01 2023].

Verkhovna Rada, 2018. *The Law of Ukraine about the Main Principles of Ensuring Cyber Security in Ukraine.* [Online] Available at: https://zakon.rada.gov.ua/ laws/show/2163-19#Text [Accessed 02 01 2023].

Verkhovna Rada, 2021. *The Law of Ukraine on the Foundations of National Resistance.* [Online] Available at: https://zakon.rada.gov.ua/laws/show/en/1702-20?lang=en#Text [Accessed 01 01 2023].

Von Tunzelmann, A., 2022. *The big idea: can social media change the course of war?* [Online] Available at: https://www.theguardian.com/books/2022/apr/25/the-big-idea-can-social-media-change-the-course-of-war [Accessed 20 12 2022].

Wahlstrom, A., Revelli, A., Riddell, S., Mainor, D., & Serabian, R. 2022. *The IO offensive: Information operations durrounding the Russian invasion of Ukraine.* [Online] Available at: https://www.mandiant.com/ resources/blog/information-operations-surrounding-ukraine [Accessed 04 01 2023].

Waltzer, M., 1977. *Just and Unjust Wars.* New York: Basic Books.

Washington Post, The, 2022. *Opinion. The Pentagon's alleged secret social media operations demand a reckoning.* [Online] Available at: https://www.washingtonpost.com/opinions/2022/09/20/military-pentagon-fake-social-media/ [Accessed 03 01 2022].

Wesolowski, K., 2022. *Fake news further fogs Russia's war on Ukraine.* [Online] Available at: https://www.dw.com/en/fact-check-fake-news-thrives-amid-russia-ukraine-war/a-61477502 [Accessed 29 12 2022].

Westby, J.R., 2011. A Call for Geo-Cyber Stability. In: *The Quest of Cyber Peace.* s.l.: ITU.

Wilmshurst, E., 2022. *Ukraine: Debunking Russia's legal justifications.* [Online] Available at: https://www.chathamhouse.org/2022/02/ukraine-debunking-russias-legal-justifications [Accessed 06 01 2023].

Wolfsfeld, G., 2022. *Making Sense of Media and Politics.* 1st ed. New York: Routledge.

Wood, M., 2022. *Everything is Possible.* New York: Holland Publishing.

Yeung, D. & Oliker, O., 2015. *Loose clicks sink ships.* [Online] Available at: https://www.usnews.com/opinion/blogs/world-report/2015/08/14/when-social-media-meets-military-intelligence [Accessed 05 01 2023].

Zakir-Hussain, M., 2022. *Russia "targeting Ukraine's communication infrastructure so they can't access news" UK says.* [Online] Available at: https://www.independent.co.uk/news/world/europe/russia-ukraine-communication-infrastructure-uk-b2029927.html [Accessed 21 12 2022].

4   Memetic War

Resistance and Actors

Tine Munk

Social Groups and Social Identity

Ukraine has employed various innovative approaches from the beginning of the war to generate substantial governmental and civic support internally and externally. These initiatives are part of the ROC concept of national resilience during peacetime, leading to resistance during conflict (see Chapters 2 and 3). While the Russian military was considered a significant power, the Armed Forces of Ukraine were highly trained but lacked the same status. However, Ukraine has proved this assumption wrong. The changes to the Armed Forces structure and command line have proved beneficial compared to the former strict hierarchical command-and-control structure. This has made the Ukrainian defence more flexible and agile, while the Russian army appears old fashioned and outdated. A notable strength of the Ukrainian system is the constitutional provision that allows the country to activate a public resistance force (Herszenhorn & McLeary, 2022; Shuster & Bergengruen, 2022). This approach was evident in the first few days after the invasion, where private support rallied internally and externally to help military units online and offline.

Some people found that the Ukrainian resilience and response surprised them, as the country has often been portrayed as being divided into two halves, the so-called "two Ukraines" (Riabchuk, 2015; Onuch & Hale, 2022, p. 21). However, this perception does not support the unity of Ukraine against the Russian invasion, which has been evident throughout 2022. While there may be extremes on both sides of the eastern and western parts of the country, Ukraine is not politically divided into two polarised parts. This is an old-fashioned mis-conceptualisation of the population. Since independence, the state has grown with a profound national sense of civic engagement, which has been instrumental in how citizens have rallied against a common enemy, Russia (see Chapter 2) (American Institute of Physics, 2022; Onuch & Hale, 2022, pp. 20–12; Zanin & Martínez, 2022, p. 7).

DOI: 10.4324/9781003432630-4

Togetherness

Human beings are inherently social creatures with a fundamental, evolutionary-based need for interpersonal relationships. Social identity and understanding peace and conflict are crucial aspects of our lives. In today's world, conflicts often arise due to differences in cultural, religious, ethnic, political, and national affiliations, which underpin most of the conflicts worldwide (Haji et al., 2016, pp. xv, xvi). Social groups shape individuals' beliefs, values, and behaviours. In-groups provide individuals with a sense of belonging, shared identity, and norms and values that define the group's culture (Hogg, 2016, p. 6).

Understanding the dynamics of in-groups and out-groups is essential for promoting social harmony and reducing intergroup conflict. In-groups are characterised by a sense of solidarity, loyalty, and trust among members. By the same token, out-groups are perceived as different and potentially threatening to the in-group's identity and values. This can lead to intergroup conflict and prejudice. Social groups, whether large demographics or smaller task-oriented groups provide their members with values and emotional identities that define them, their beliefs, and their behaviours (Hogg, 2016, p. 6). Therefore, effective communication from both authorities and citizens is essential to boost morale and re-emphasise shared identity and cultural belonging, especially in the face of an existential threat from an out-group. It is important to note that in-groups have a different social context from out-groups. The warmongering from the Russian side promotes a pseudohistorical outline of shared history and the claim of being "one nation". However, this argument is counterproductive as it only creates a more potent shared identity in Ukraine and boosts the patriotic credentials of politicians in power (Molchanov, 2016, p. 202; Fiala, 2022).

Ukraine has faced numerous challenges and threats to its sovereignty over the past several years, including the Orange Revolution, Euromaidan Revolution, and the Russian hybrid war, where civic engagement has been a critical factor. However, ROC's resilience and resistance approach involves mobilising civil society groups and volunteers to support the Armed Forces, governmental agencies, and actors (see Chapter 2). The response to the conflict in Ukraine has also challenged the propaganda claims made by Russia that Ukraine has no agency and is not unified. The high levels of civic engagement and resistance have demonstrated the strength and resilience of the Ukrainian people, with 80% of the population involved in the defence, 45% donating money, 35% volunteering, and 18% taking part in information resistance since the 2022 war began (Rating Group, 2022; Zarembo, 2022). Through voluntary initiatives, groups and individuals provide essential aid and support to those affected. These efforts are instrumental in ensuring the country's continued functioning amidst the challenges posed by the ongoing war. The initiative spans setting up local community support networks, organising humanitarian assistance, and crowdfunding for the military. The Ukrainian society has a flexible and agile approach to civic defence and the protection of citizens (Onuch & Hale, 2022, pp. 14–15).

The war in Ukraine came as a shock to many people, both internally and externally, highlighting the impact of war on ordinary life and civil populations. The use of live streaming, photos, and eyewitness statements has exposed the damage caused by the conflict and has helped garner support for the Ukrainian side. Humour and satire have played a critical role in creating a sense of solidarity, easing fear and uncertainty, and releasing tension during these traumatic times (Sirikupt, 2022). Ukrainian humour is a helpful tool for managing anger and frustration, even when it becomes dark and sharp – Ukrainians joke about the defeat of Russian soldiers, the impact of sanctions, and propaganda absurdity. Memes also focus on everyday life and how Ukraine manages the war situation. Meme communications cover power outages, Russian attacks on critical infrastructure, sheltering, air sirens, and missile attacks (Antoniuk, 2022). Memes can serve as a front for more severe objections to Russian aggression. For example, memesters have used the image of the Russian military and President Putin in various ways while simultaneously cultivating solidarity among civil supporters, internal and external to Ukraine (see Chapter 5) (Sirikupt, 2022)

Different Types of Actor

Governmental and non-governmental actors are equally essential to the progress of memetic warfare. Political actors are primarily defined as those seeking to influence decision making through organisational and institutional means. The category of political actors is broad and is most likely linked to established political parties or like-minded individuals working together within an agreed organisational and ideological structure, where they join forces to pursue a common goal (McNair, 2011, p. 5). Debunking and using public information circulated on social media and virtual spaces is one way the government has filled in some of the existing information vacuums (OECD, 2022).

Ukraine has adopted cybercrime legislation and cybersecurity strategies to regulate the online environment. Still, memetic activities predominantly operate within the law without being formally recognised as part of the information war strategy. Nevertheless, it is an area where some overlaps with information and conventional warfare can cause problems. Without a clear strategy and tactics for controlling and progressing memetic data, there is a risk of losing vital information when different actors operate in independent and relatively closed networks. A more formal strategy with defined roles and communication routes, including a form of ROC resilience structure, would help manage the information collected during memetic warfare (see Chapters 1 and 3). Different types of communication would help continue to debunk information from Russia and target the adversaries based on their online engagement and communications.

There appears to be a complex web of actors and motivations involved in memetic warfare, which includes both state and non-state actors. Understanding the diverse range of actors involved in memetic warfare is essential to develop effective strategies for countering its effects. Social media have significantly generated large-scale support for Ukraine through digital environments and online communities (Bernot & Childs, 2022). Online resistance involves various groupings and motivations, including war and warfare, independent militant or patriotic actions, civil disobedience, and offline and online activism (Munk, 2022, p. 20). Numerous groups and individuals are involved in memetic warfare, including military combatants and civilians. Private actors are also employed for their expertise, alliances, and support, often without being part of the state apparatus. Other actors can be individuals or groups hired for specific jobs or sponsored by the state without being a direct part of the state apparatus or through delegation of authority (Holt & Bossler, 2016; Munk, 2022, p. 26). Memesters and non-skilled actors collaborate with IT and computer-savvy individuals to protect against hackers, online interference, propaganda, and dis-and-misinformation. These civic volunteers have carved out a space in the resistance by working with national and international companies and various governmental actors (Bernot & Childs, 2022).

Governmental Actors

The importance of social media for the Ukrainian government in collecting and spreading information to its population and lobbying for international support has increased significantly due to the ongoing war and Russian aggression. The urgency of countering these actions has highlighted the need to use social media platforms actively. The government, under the leadership of President Zelenskyy, is confident in Ukraine's potential to use online and offline communication strategies to win friends and support. The official communication from the authorities is inclusive and targets a wide range of groups and individuals within and beyond Ukrainian citizens. The communication action plan includes a media strategy that promotes the nation, culture, and ongoing defence of the country. This strategy is aimed at engaging people internally and externally and strengthening Ukraine's position in the global community (Ellwood, 2022; OECD, 2022).

The Ukrainian authorities have recognised the importance of using videos and images as a military means to achieve their political goals, a concept known as image warfare. President Zelenskyy, an effective communicator, plays a crucial role in this strategy, using his speeches and video clips to spread knowledge and appeal for help and support from the Western world and the natural allies of Ukraine (see Chapter 5). The Ukrainian authorities are comfortable using digital tools and recognise the advantages of social media platforms for communication (Benebid, 2022, p. 4; OECD, 2022). Key actors, such as the Commander-in-Chief of the Armed Forces of Ukraine, Zaluzhnyi, and the Ukrainian Defence Minister, Reznikov, regularly post short messages and videos on Twitter and other online channels. Other ministries, governmental actors, and Armed Forces personnel follow the same communication strategy, providing the public with online communications. While these communications may sometimes appear coordinated, they are often linked to the actual person, unit, or department to provide essential information to the public. The Armed Forces of Ukraine and the General Staff of the Armed Forces of Ukraine have a significant online presence, sharing information about the war situation and progress, despite the challenges associated with national security (see Figures 4.1, 4.2) (Defence of Ukraine, 2012; General Staff of the Armed Forces of Ukraine, 2015; Oleksii Reznikov, 2015;

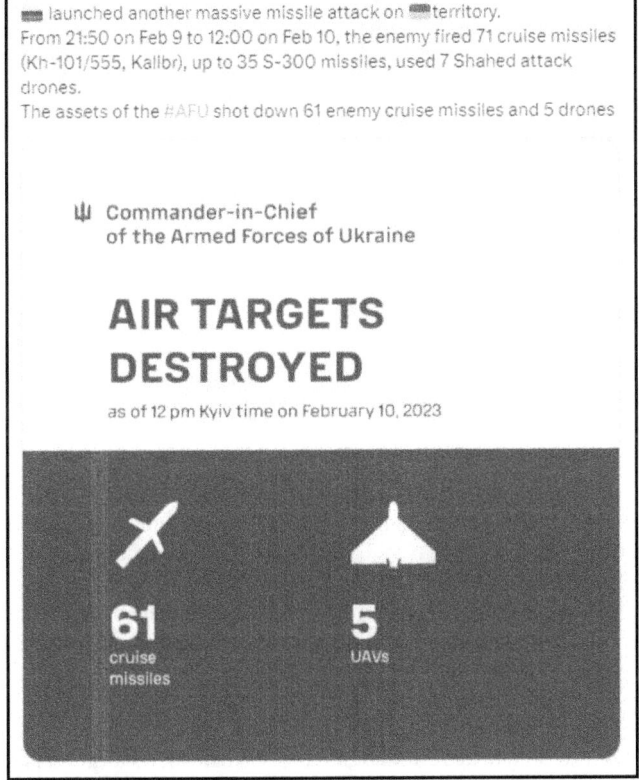

Figure 4.1 Communication from the Commander-in-Chief of the Armed Forces of Ukraine

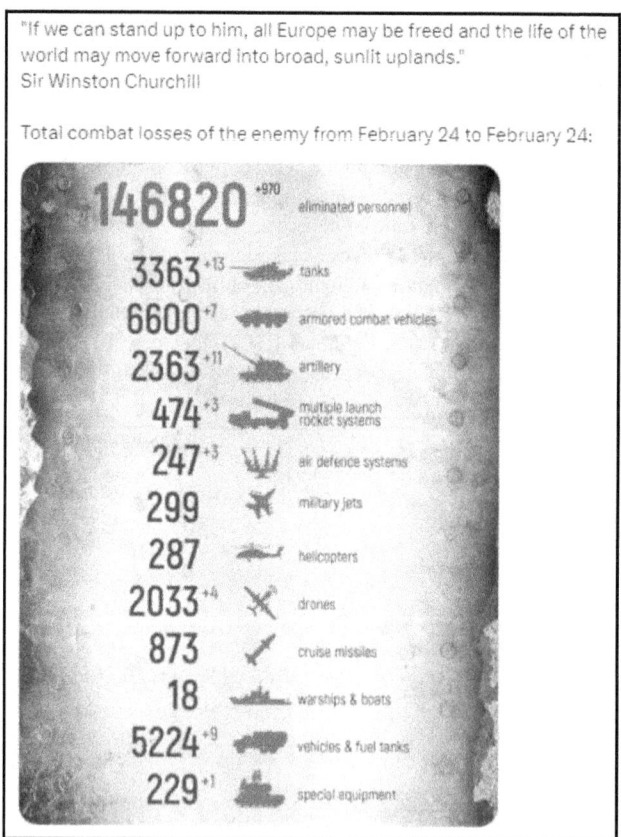

"If we can stand up to him, all Europe may be freed and the life of the world may move forward into broad, sunlit uplands."
Sir Winston Churchill

Total combat losses of the enemy from February 24 to February 24:

146820 +970   eliminated personnel

3363 +13   tanks

6600 +7   armored combat vehicles

2363 +11   artillery

474 +3   multiple launch rocket systems

247 +3   air defence systems

299   military jets

287   helicopters

2033 +4   drones

873   cruise missiles

18   warships & boats

5224 +9   vehicles & fuel tanks

229 +1   special equipment

Figure 4.2 Communication from the Defence of Ukraine

Commander-in-Chief of the Armed Forces of Ukraine, 2022; Benebid, 2022; OECD, 2022).

The conflict in Ukraine has been kept alive in the public sphere due to the constant flow of information and communication facilitated by digital media. This has made it easier for news and updates to spread quickly, thereby keeping the conflict in the public eye and enabling the Ukrainian government to maintain its relevance and garner support from various quarters. The Minister for Digital Transformation of Ukraine, Fedorov, has been leveraging digital technology to communicate with businesses and individuals locally and globally to promote the government's political agenda and seek support. He has constantly communicated with businesses online, asking for help and trying to

Did you know that European, American, British companies still make $700m daily by trading with Russia? Write to an international company of your choice: oil and gas, car manufacturers, investment funds, retailers etc. Demand them to cut all ties with Russia. More: @Boycott_RU

CORPORATIONS MUST STOP FINANCING PUTIN'S WAR.

BOYCOTT RUSSIA

Figure 4.3 Call for Support

persuade them to withdraw from Russia (see Figure 4.3). Fedorov has also directly contacted companies such as Asus, Amazon, PayPal, Microsoft, Apple, Google, and NETSCOT to request their assistance (Bernot & Childs, 2022; OECD, 2022).

It is interesting to see how the Ukrainian government has been leveraging digital technology to communicate with businesses and individuals locally and globally to promote its political agenda and seek support. The interaction with Elon Musk is a prime example of how such efforts can yield positive outcomes. On Twitter, Fedorov asked Musk for access to SpaceX's Starlink system. Musk responded with a tweet confirming that "Starlink was active" in Ukraine. This system replaced the Internet infrastructure Russia had destroyed (Bachman, 2022; Bernot & Childs, 2022; OECD, 2022).

Official Twitter Accounts

The official Twitter accounts of Ukrainian government departments and the Armed Forces are essential communication channels. These accounts feature diverse content, including images and videos, as part of the country's wartime strategy. Since the full-scale invasion in February 2022, the tone of official posts on Twitter has evolved. However, there is still a mix of humour and serious information about Ukraine's current situation. While official communicators using these accounts aim to balance providing relief and delivering relevant updates, they are mindful of avoiding jokes that might be insensitive to the suffering and fear experienced by Ukrainians during the war. Finding this balance is particularly challenging in times of war, as officials strive not to upset the population or jeopardise the support they receive from within and outside the country. Nonetheless, even during bleak moments, people need

Figure 4.4 Official Twitter Account for MFA of Ukraine

Figure 4.5 Official Twitter Account of Ukraine

Figure 4.6 Official Twitter Account for Defence of Ukraine

moments of levity and a sense of solidarity. Memes circulated on these official Twitter accounts are often dark and witty, reflecting the Ukrainian spirit and using irony and sarcasm to convey actual events (Dominauskaitė & Svidraitė, 2022; Ukraine/Україна, 2022). The Twitter content includes jokes, war updates, memes, and symbols of Ukraine, such as the flag, trident, watermelons, sunflowers, and messages from President Zelenskyy and other officials (see Chapter 5) (Ukraine, 2016; Snowden, 2022).

Most communications, such as speeches, memes, and videos, are accompanied by subtitles or produced in English (see Figures 4.4–4.6). Using an international language allows Ukrainian state actors to garner more followers and support from the global community (OECD, 2022; Sirikupt, 2022). This enables online resistance groups to utilise their trolling skills and to create and disseminate anti-Russia humour. Pro-Ukraine users often respond to these communications in the comment section, where spin-off memes are uploaded, taking on new shapes and forms that ensure a longer online life (Sirikupt, 2022). For example, the video version of tractors towing away Russian tanks and military equipment under the title "Do not mess with Ukrainian farmers" posted by the Official Account of Ukraine, @Ukraine, elicited supportive videos and comments from online users, along with the Ukrainian flag and hashtags such as #SlavaUkraini (Glory to Ukraine) and #StandWithUkraine in solidarity (see Chapter 5) (Sirikupt, 2022; Ukraine/ Україна, 2022).

Figure 4.7 Ukrainian Farmers' Army (a)

Figure 4.8 Ukrainian Farmers' Army (b)

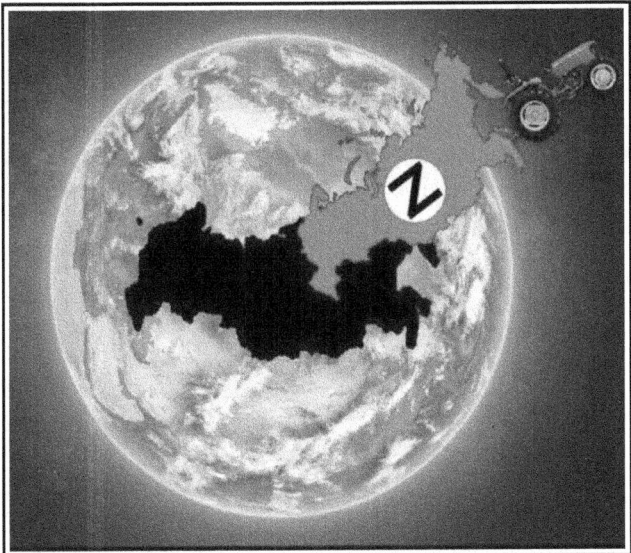

Figure 4.9 Ukrainian Farmers' Army (c)

Videos

Video-memes are rapidly disseminated and remixed messages circulated among members of a particular political or social culture for purposes such as satire, parody, or critique (Mortensen & Neumayer, 2021, p. 2369). Memes that includes short videos have become increasingly ingrained in resistance efforts as a part of information warfare and government communication strategies, serving both informative and humorous purposes while also fostering a sense of internal solidarity in opposition to an "us vs. them" mentality.

Crimea Video

In August 2022, an attack was launched against the Saky airbase in Crimea. Russian tourists who were on holiday in the peninsula recorded the explosions from the airbase, and soon after, they were seen fleeing the island. The tourists recorded themselves in a state of shock and despair as they left what they thought was a safe holiday destination, only to be caught in a massive traffic jam while crossing the Kerch bridge (Bickerton, 2022; Sabbagh, 2022). One particular video of a crying Russian tourist went viral as she was recorded saying: "I don't want to leave Crimea. It's so cool here and it's like being at home" (Bickerton, 2022).

This clip became part of memetic warfare, in which counter-memes ridiculed Russians holidaying in Crimea during a war. The Kyiv government did not take responsibility for the attacks. An advisor to President Zelenskyy suggested that the attack might have been carried out by partisans operating behind Russian lines in Crimea (Sabbagh & Lock, 2022). Russian authorities attempted to downplay the attacks by falsely claiming that the Saky airbase attack was an accident caused by several aviation munitions detonating in a storage area, despite satellite photos showing numerous impact craters and damaged or destroyed aircraft on the base (Bickerton, 2022; Sabbagh & Lock, 2022).

The Ukrainian defence ministry quickly tweeted a link to the short 35-second video of the fleeing Russians, with captions that advised Russians to leave the country (see Figure 4.10) (Adams, 2022; Bickerton, 2022; Defence of Ukraine, 2022e; Sabbagh & Lock, 2022).

War Videos

This is not the only video the Ministry of Defence produced as a part of its memetic warfare strategy. The governmental department has been very productive in developing memes and short videos. The use of humour and satire is aimed at ridiculing the enemy and as a coping mechanism for the Ukrainian people during wartime. By making fun of the situation, the government is attempting to alleviate some of the stress and anxiety experienced by Ukrainians while also fostering a sense of national unity and solidarity (Dominauskaitė & Svidraitė, 2022).

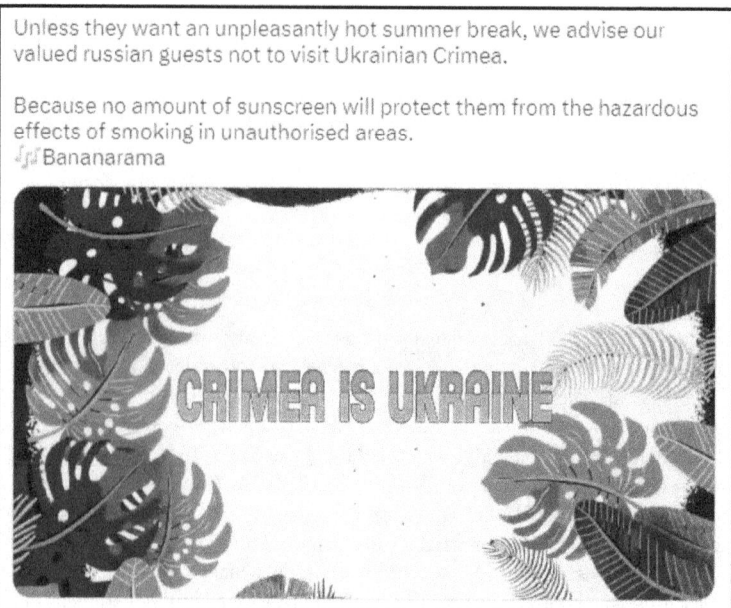

Figure 4.10 The Attack on Saky Airbase 2022

It is interesting to observe how the Ukrainian Ministry of Defence uses memes and short videos as a part of its memetic warfare strategy. By producing these videos, they can reach both an international and external audience, similar to the effect of President Zelenskyy's speeches. It is also noteworthy that small groups of private actors manage social media volunteers delivering these videos to the ministry's followers internally and externally (Adams, 2022). Demonising Russian losses are common in these videos, as seen in the "Run, Rabbit, Run" video about the Russian forces at Kherson Blast in August 2022 (Adams, 2022; Defence of Ukraine, 2022e) and the "festive" Christmas video of how the underdog wins against the bad guys (Defence of Ukraine, 2022f). This video is based on the *Die Hard I* movie from 1988, which shows how popular culture is being incorporated into the memetic warfare strategy (IMDb, 1988; Powell, 2022).

One aspect of memetic warfare is the use of online communication forms and visual images to convey a negative impression of the enemy, often through a specific humorous narrative about their actions (Screti, 2013, p. 212; Ross & Rivers, 2017, p. 3). In late 2022, Ukrainian Defence Minister Reznikov created a personal address to Santa Claus, expressing a hope for victory in the upcoming year (see Figure 4.11). The resulting memetic video presents a romantic and naive Christmas illusion of a child writing a letter to Santa, with references to the ongoing war and a sarcastic jab at Russia:

> I hope you don't mind that we generously shared some of your gifts with our deranged northern neighbour who behaved very badly this year. (Oleksii_Reznikov, 2022)

In the video, Reznikov asks Santa to deliver a present to President Putin, in the form of a defeat in the ongoing war. The video is meant to be humorous

Figure 4.11 Official Video from the Ministry of Defence (a)

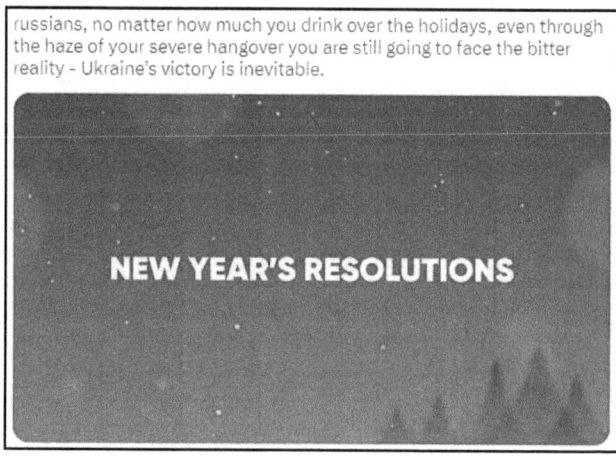

russians, no matter how much you drink over the holidays, even through the haze of your severe hangover you are still going to face the bitter reality - Ukraine's victory is inevitable.

NEW YEAR'S RESOLUTIONS

Figure 4.12 Official Video from the Ministry of Defence (b)

and playful but also ridicules the Russian leader and portrays Ukraine as the underdog fighting against a powerful aggressor. This type of memetic warfare effectively shapes public opinion and generates support for a particular cause or viewpoint.

The same governmental department sent a New Year video message to Russia that featured the defence minister talking directly to the Russians liable for military services. The video message is: "Surrender or die" (Defense of Ukraine, 2022h). Following New Year's Eve 2023, another video was posted after a night of heavy missile attacks on key cities. This video footage gives insight into the reality of the Russian invaders when they wake up after their New Year's celebrations. The video is supported by "Auld Lang Syne", a well-known song primarily sung at New Year (see Figure 4.12).

"Thank You" Videos

Videos can also serve as a positive form of communication that legitimises different activities, and creating positive links to allies is crucial. Memes allow governmental departments to legitimise their defence strategy and weapon deliveries to the military by focusing on progress in the war (Steffek, 2003; Ross & Rivers, 2017). These communications are essential to reach out and express gratitude for the weapons and support received by state officials and citizens. These types of video serve to both legitimise Ukraine's defence activities and demonstrate appreciation for the support received from allies. For instance, the Ministry of Defence's webpage features a video titled "Friends Will Be Friends", a visual "thank you" message to the UK for providing anti-tank weapons support. The video references well-known tourist and cultural events and destinations, linking the theme to James Bond and Russian villains. The British punk-rock band, The Clash, supports the visuals with upbeat music (Adams, 2022; Defence of Ukraine, 2022d). "Merci beaucoup France!" The video shows French military aircraft arriving in Ukraine with Caesar self-propelled weapons (Defence of Ukraine, 2022a). The video expresses gratitude to France, with scenes featuring the weaponry in action and a voice-over message thanking the French government and people for their support: "Romantic gestures take many forms" (Defence of Ukraine, 2022a)

This message was, of course, followed by romantic images of red roses, chocolate, and the Paris skyline. The video was accompanied by Gainsbourg and Birkin's *Je t'aime moi non-plus* to emphasise the message of true love and the key text:

Nothing beats 155mm highly mobile self-propelled artillery. (Adams, 2022; Defence of Ukraine, 2022a; Srivastava et al., 2022)

Similarly, Sweden was also acknowledged with a "thank you" video that included footage of the Carl Gustav rocket launcher, which can take out a Russian T-90 tank worth $4.5m. The Defence Ministry used ABBA's *Money, money, money* as the underlying music in the video (Adams, 2022; Defence of Ukraine, 2022c). Norway was honoured with a beautiful winter-themed "thank you" video for providing NASAMS to protect Ukrainian cities. The video references Norway as "the land of the Peace Prize", with the underlying music being *Peer Gynt, op. 46, IV*, by the Norwegian composer Grieg (Defence of Ukraine, 2022g). Countries including Germany, Denmark, and Finland, have also received a short video thanking them for their help (see Figures 4.13–4.15) (Defence of Ukraine, 2022b).

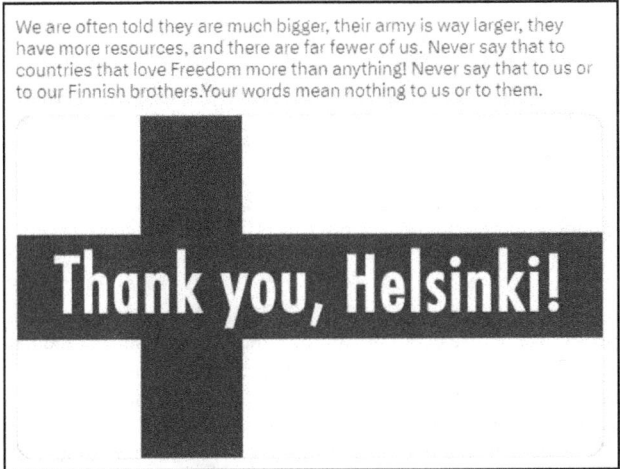

Figure 4.13 Thank You Video: Finland

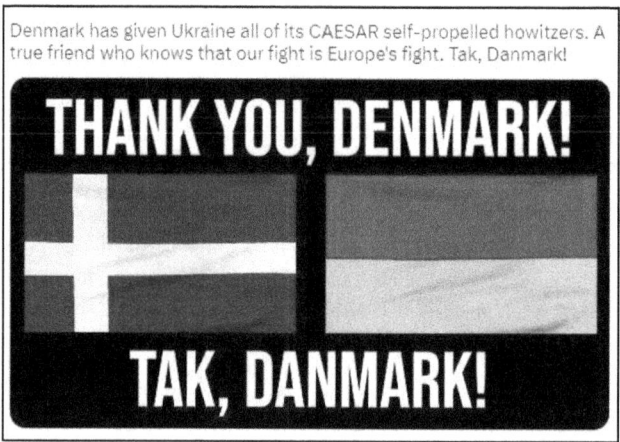

Figure 4.14 Thank You Video: Denmark

Figure 4.15 Thank You Video: Germany

The Cyber Army

Two days after the invasion, Ukraine's Deputy Prime Minister and Minister for Digital Transformation, Fedorov, launched an IT Army based on volunteers (see Figure 4.16) (Anton, 2022; Burgess, 2022; Fedorov, 2022; Munk & Ahmad, 2022, p. 231; Serrano, 2022). While the initial call was for digital

We are creating an IT army. We need digital talents. All operational tasks will be given here: t.me/itarmyofurraine. There will be tasks for everyone. We continue to fight on the cyber front. The first task is on the channel for cyber specialists.

t.me
Telegram: Contact @itarmyofurraine

Figure 4.16 Call for the IT Army

talent to engage in cyberwarfare, it was later extended to everyone (Burgess, 2022; Serrano, 2022). Online users, both from Ukraine and abroad, can sign up for this army through a Telegram channel, and people are selected based on their skills, such as developers, cyber specialists, designers, copywriters, and marketers, to enhance defence (Ministry of Digital Transformation of Ukraine, 2022; Zaliznyak, 2022). This call for action is positioned at the fringe of memetic warfare; however, some activities overlap (Zaliznyak, 2022). As part of ROC, which advocates for the merger of resistance tactics between public and private grounds included in warfare, authorisation from an established authority can give the IT Army credibility and separate it from other civic elements of the resistance (Maskaliūnaitė, 2021, p. 33). However, concerns have been raised about external IT Army participants engaging in activities that are criminalised in their home country, such as hacking. Such actors would break national laws and risk prosecution if they engage in hacking (Munk & Ahmad, 2022, pp. 23).

Thousands of volunteers have joined the IT Army, but little information has been published about the actors and their level of engagement. The Telegram channel has shared screenshots of websites that have been taken offline, but it is unclear where these images originated from or the success rate of these actions (Burgess, 2022; Griffin, 2022). Tasks have been launched for less tech-savvy members, such as reporting Russian YouTube channels spreading misinformation about the war. These channels include state-owned media outlets such as Russia 24, TASS, and Ria Novosti, or directly debunking disinformation online, which shares similarities with memetic warfare (Anton, 2022; Griffin, 2022; Serrano, 2022). The IT Army has also assisted with the cyber defence of key Ukrainian institutions to ensure that their functions and services are available to citizens. These institutions include the National Bank and the Cabinet of Ministers (Cain, 2022).

The Private Actors

The threat to cybersecurity also affects the private sphere, where individuals act independently of military units or political causes and promote violence and harm to spread political, societal, or ideological fear (Parker & Kremling, 2017, p. 89; Munk, 2022, p. 32). Volunteering in support of the army is deeply embedded in Ukraine's civic society, and has been since 2014 when volunteers contributed to fundraising and collecting items such as boots and field hospitals (Yekelchyk, 2020, p. 135). However, this concept was reinforced in the 2022 war, where a widely circulated online post called for volunteers on social media. Governmental and non-governmental organisations such as Saint Javelin, United24, Dzyga's Paw, and the Georgian Legion mainly operate in supporting and fundraising (Dzyga's Paw, 2022; Saint Javelin, 2022; United24, 2022b). As the Russian invasion continues, there has been a significant call for support outside Ukraine to assist in the different efforts to aid Ukrainians. Communities are coordinating donations for Ukrainian refugees and initiatives launched within the country. The online environment has facilitated the spread of information about these donation options and charities (Pantic, 2022). The meme culture has been beneficial for these organisations to engage a broader community network with combinations of informative online posts or memes on leading social media sites.

Voluntary engagement in the Ukrainian resistance has taken another active turn, which is unprecedented in modern warfare. Private actors have been actively countering Russian propaganda, which was previously seen as a "mission impossible". There are several Twitter communities, including meme groups such as NAFO (North Atlantic Fella Organization, 2022), Ukraine Memes for NATO Teens (Ukraine Memes for NATO Teens, 2022), and the Ukrainian Meme Force (UMF) (Ukrainian Meme Force, 2022), along with several other meme-centred non-governmental campaigns flourishing on social media in support of Ukraine (Shultz, 2022). The main aim of these groups is to boost the Ukrainian online counter-propaganda effort (Bishara, 2022). Unlike NAFO, which is a global private voluntary community endorsed by many politicians and state officials worldwide, the UMF was created as a semi-official arm of the Ukrainian civic resistance.

Hacktivism

Hacktivist groups have also joined the fight against Russia on the new memetic frontier. While most hacktivist groups, such as Anonymous, are known for direct cyberattacks, their diversified hacking strategy now includes memes. Over the years, these groups have developed a disruptive and effective campaign of modern crowdsourced cyberwarfare (Pitrelli, 2022). Hacktivists use legal and/or illegal digital tools to achieve political goals, such as promoting free speech and human rights, in an online environment linked

Figure 4.17 Group Anonymous' Engagement with the War

to traditional political protest or civil disobedience (Jordan & Taylor, 2004, pp. 68–69; Yar & Steinmetz, 2019, pp. 85–86; Lavorgna, 2020, p. 60; Munk, 2022, p. 32). After declaring war on Russia, Anonymous hacked into state-owned television and changed the content to inform about the war in Ukraine, as Kremlin controls the narrative and censors the news flow (see Figure 4.17) (Munk & Ahmad, 2022, p. 232; Nicholson, 2022). Anonymous operates by launching "wars" against corporations, states, and individuals who disagree with or violate the group's moral code (Munk, 2022, pp. 214–215). Group members have previously used memes to ridicule and debunk the terrorist group ISIS's activities. Now, the group's memetic warfare is directed towards Russia and Putin. The operations combine hacking missions and uploading memes about their activities and opinions on social media (Mortensen & Neumayer, 2021; Sirikupt, 2022).

One of Anonymous' earliest actions involved hacking into marine tracking data and changing the call sign of Putin's yacht to "FCKPTN" and the destination to "Hell" (see Figure 4.18) (Munk & Ahmad, 2022, p. 233; Teh, 2022) Memes mocking Putin and his assets emerged from this action in the online environment, utilising the power of memetic warfare in addition to more conventional hacktivist tactics.

Memes such as these have been influential by creating a playful atmosphere linked to sharing identity and a feeling of togetherness in an online battle against an identified enemy. This has pushed the online frontier where memetic communications are equally crucial to the hack. The use of humour, irony and sarcasm in these memes overshadow the illegal act of getting access to maritime data (Mortensen & Neumayer, 2021). The use of memetic warfare can potentially engage a broader audience and shift the narrative in favour of hacktivists. In a time when social media are a primary source of news for many, memes can impact online users' opinions. This is why hacktivist groups including Anonymous have embraced memetic warfare as a powerful tool to fight against their enemies.

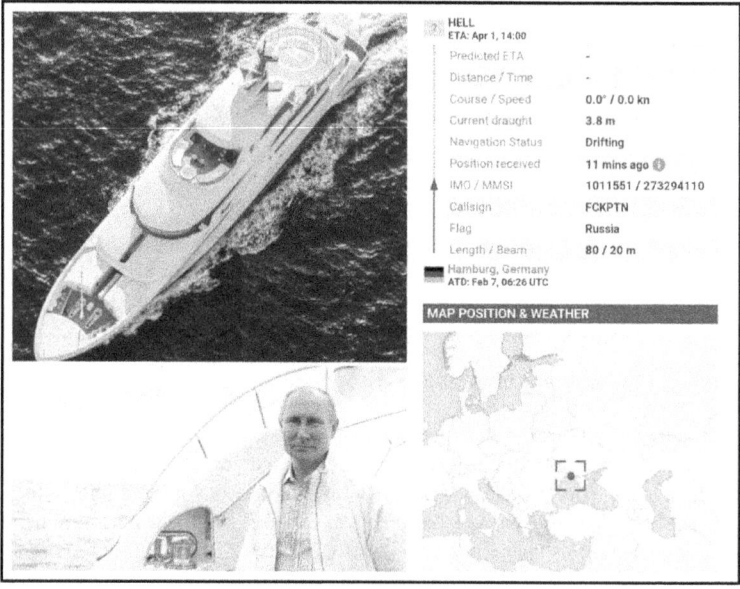

Figure 4.18 Anonymous' Hack

Go to Google Maps. Go to Russia. Find a restaurant or business and write a review. When you write the review explain what is happening in Ukraine.
Idea via ████████████

Figure 4.19 Anonymous' Invitation to Inform about the Situation in Ukraine

The hacktivist group has also been successful in mobilising a larger group of online users to counter misinformation about the war in Russia. Anonymous called for its followers to post information about "what is happening in Ukraine" to counter Russian propaganda and dis-and-misinformation (Munk & Ahmad, 2022). Anonymous called for followers to comment on the Ukraine situation by posting information about "what is happening in Ukraine" on Google Maps, Afisha.ru, and TripAdvisor (see Figure 4.19). This action led to a flood of social media posts, which helped to raise awareness and provide alternative perspectives on the conflict.

This Anonymous-supported action aimed to circumvent Russian censorship while explaining to Russian online users what was happening in Ukraine (see Figure 4.20) (Deighton, 2022; Gronholt-Pedersen, 2022; Munk & Ahmad, 2022, p. 234; Smith, 2022). This action led to suspending reviews on Alphabet Inc.'s Google Maps and travel platform TripAdvisor Inc. to prevent Ukraine supporters from posting more political statements on businesses and tourist destinations pages.

Translation: The food was great! Unfortunately, Putin spoiled our
appetites by invading Ukraine. Stand up to your dictator, stop killing
innocent people! Your government is lying to you. Get up!

Figure 4.20 Example from the TripAdvisor Campaign

Figure 4.21 North Atlantic Fella Organization Logo

What is #NAFO? NAFO is the HIMARS of social media.
#NAFOfellas use the most powerful weapons: (un)sophisticated memes
& and satire to hurt the feelings of sensitive Russian trolls. And yes, we
will win the info war.

Figure 4.22 NAFO Definition

#NAFO

There is also space for more innovative forms of collective resistance in the
war. Online memes, counting disinformation and fundraising, are taking
many forms as a part of hybrid warfare. There is a high level of creativity in
the actions to take on the Russian propaganda machine (North Atlantic Fella
Organization, 2022; Ukrainer, 2022). However, one of the more innovative
groups is the Fellas – also known as the North Atlantic Fella Organization
(NAFO), which plays on the acronym of the North Atlantic Treaty Organization, i.e. NATO) (see Figures 4.21, 4.22). NAFO started in May 2022, when
the Polish Twitter memester Kama Kamelia began creating avatars for Twitter
users (Fellas) based on a binary Japanese Shiba Inu dog, often in a military
uniform or with links to Ukraine. These avatar dogs are dressed up at the do-
nor's request. Originally the pro-Ukrainian Fella avatars were given to online
users who donated to the Georgian Legion of the Armed Forces of Ukraine.
However, this approach is expanded to all types of donation to Ukraine that
can be verified (Ukraine, 2016; McInnis et al., 2022; Taylor, 2022).

The NAFO members' meme creations are constantly spreading and changing across the political spectrum and platforms. These memes have their existence, and the memetic visuals become remixed, iterated and decontextualised. The memes and the actors are intertwined in a political context where a participatory Internet culture becomes an important player in Ukrainian memetic warfare (Mortensen & Neumayer, 2021, p. 2373).

NAFO is an unofficial frontline army known as "shitposters", where members have formed an informal community on Twitter to communicate about Ukraine, trolling accounts to counter Russian dis-and-misinformation, and raise money for Ukraine. The group uses memes and collages, often featuring the avatar dogs, to ridicule and challenge pro-Russian accounts (see Figures 4.23–4.25). The memes use a particular language, including Russian

Figure 4.23 NAFO Meme (a)

Figure 4.24 NAFO Meme (b)

Figure 4.25 NAFO Meme (c)

words such as "vatnik" for a Russian/Kremlin supporter, or such phrases as "What air defence doing?" (see Figure 4.24) (Michaels, 2022). Most of the communications are in English. However, NAFO's use of poor English is a deliberate tactic to mock the language skills of Russian supporters. This kind of trolling war is reminiscent of the tactics used by Russian troll farms in Saint Petersburg (Taylor, 2022; Ukrainer, 2022).

NAFO is not a military alliance as NATO is, but the members react to the same core principle included in Article 5. If a NAFO member receives threats or finds Russian propaganda on Twitter that requires joint action, they can activate Article 5 in a tweet. The members will respond like a pack of dogs and flood the target's comment field with numerous memes and posts. NAFO is not aiming to discuss the dis-and-misinformation; instead, they are countering them online using absurd language and images in open mockery (see Figures 4.23, 4.25). The memes launched by NAFO are often so ridiculous that they are impossible to engage with or counter. This is the core element of NAFO's strategy, and it is successfully confusing its targets (Braun, 2022). Many Russian or Russian-supporting online users are feeling the effects of the organisation's actions and constant online presence. They often express annoyance and confusion with NAFO's activities, which include overloading comment fields with memes, getting accounts closed, and uploading facts. NAFO's targets struggle to understand the organisation and management of the group and argue that it is a bot or is developed and paid for by NATO and the CIA (see Figures 4.26, 4.27).

The Ukrainian government has been eager to highlight the memes from NAFO on its social media webpages, expressing gratitude with a salute. The

> The intel community usually has strict rules not to use social media. But for the first time the entire global workforce of US & partner spy agencies has permission to engage the enemy on Twitter via #NAFO using aliases and fake accounts combined with bots. Social media warfare.

Figure 4.26 Reactions to NAFO Actions (a)

> NAFO is a CIA op.
>
> Anyone that willfully supports a genocidal, warmongering institution like NATO must be a dupe of the intelligence services or a bot, full stop.

Figure 4.27 Reactions to NAFO Actions (b)

community is systematically targeting official pages of Russian state institutions and followers. Members are openly harassing those pushing Russian propaganda online, writing mocking replies, creating and sharing memes with their Fellas, and reporting the pages to social media moderators (Ukrainer, 2022). NAFO has been particularly successful in reporting bots and propaganda to social media companies, and has helped identify these accounts so they can be suspended or locked (McInnis et al., 2022).

NAFO has grown from one online user having fun on Twitter to a global phenomenon, with several politicians becoming honorary members. The group is not linked to any organisations or institutions but operates as a voluntary, meme-empowered community of like-minded people who want to support Ukraine (Ukrainer, 2022; Kirichenko, 2023a). The initiatives and the work of the Fellas in the memetic war are praised by world politicians (Drummond, 2022). The Ukrainian Minister of Defence, Reznikov, honoured the NAFO membership by changing his profile picture to his NAFO dog avatar for a few days (see Figure 4.29). Estonia's Prime Minister Kallas also recognised her membership, and her personalised Fella was sporting a yellow

Figure 4.28 Defence of Ukraine's Salute to the Fellas

> My personal salute to #NAFOfellas. I'd like to thank each person behind
> Shiba Inu cartoon. Your donations to support our defenders, your fight
> VS misinformation is valuable.
> I'm changing my profile picture for a few days. Cheers ████████
> NAFO expansion is non-negotiatiable!

Figure 4.29 Ukrainian Ministry of Defence Oleksii Reznikov's Salute to the Fellas

> My greetings to #NAFOfellas - you're doing a great job fighting bad takes
> and Russian propaganda, and raising funds for Ukraine's defence. I
> salute you, #Fellas.
>
> Thank you for gifting me my own Fella, @Phoenixfire709. #NAFO
> expansion is non-negotiable.

Figure 4.30 Estonia's PM Kaja Kallas' Greeting to the Fellas

blazer and a blue shirt (see Figure 4.30) (Michaels, 2022; Ukrainer, 2022; Kirichenko, 2023b)

For the collective to continue, togetherness and recognition are important factors; the politics of inclusion and exclusion are embedded in the collective, enabling the growth of the resilience movement and a feeling of being on a joint mission supporting Ukraine. Finally, having fun while "bonking" should also be factored into the defensive action and engagement of the Fellas.

The community feeling is strong within the collective. Although everyone can become a part of this group, the joint actions and the interactions between the members have created bonds across social, economic, political and cultural differences. Following other Fellas online is instrumental in developing vast social NAFO networks internally in the group (see Figure 4.34). The NAFO community is focused on its mission to support Ukraine, combat Russian propaganda, and take care of one another's well-being. The members of NAFO are concerned not only with the larger political issues at hand but also with their fellow members' mental and emotional well-being. The actions of the Fellas place them on the online frontline of the Russian propaganda machine. This information can have a negative impact on the Fellas, who interact with and encounter some horrifying memes, comments, and videos online during these activities. It is common to see a NAFO hug and support memes circulated on Twitter with a NAFO dog and a kind or inspirational quote/text (see Figures 4.31–4.33). Recognising the potential negative impact of engaging with such content online and the support provided to members struggling with it is a testament to the strong community bond within the group.

Figure 4.31 NAFO Support (a)

Figure 4.32 NAFO Support (b)

Figure 4.33 NAFO Support (c)

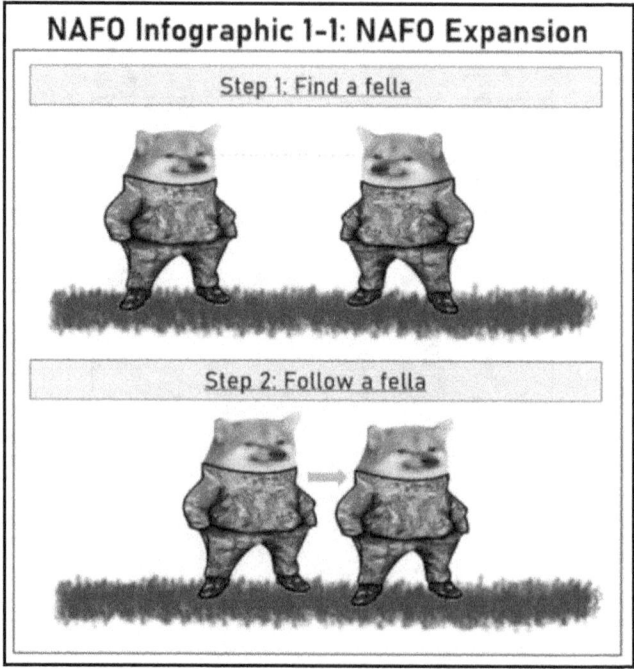

Figure 4.34 NAFO Support (d)

Memetic Weapons and Attacks

Internet memes have emerged as a powerful tool in the Ukrainian resistance against Russian aggression and propaganda. NAFO memes serve to boost morale, unify Ukrainian civil society, ridicule the Russians, highlight Russian aggressions, and inform the international community about events on the ground (Ukrainer, 2022). However, the NAFO Fellas are also known for their combative attitude and desire to engage with Russian officials through social media memes and commentary. Ulyanov, the Russian diplomat to the International Organizations in Vienna, experienced the power of NAFO when he attempted to engage with a member who mocked him online (Drummond, 2022; Taylor, 2022; Kirichenko, 2023a, 2023b) During the exchange, Ulyanov blamed Ukraine for Moscow's invasion and shelling of civilians. When he tried to counter a NAFO Fella's mocking comments, he tweeted: "You pronounced this nonsense. Not me" (Mikhail_Ulyanov, 2022).

Ulyanov was unaware of what he was up against when he began communicating with the Fellas online, and he found himself in a regular "shitstorm". The reaction came instantly from the NAFO community. When Russian governmental actors or supporters engage with the NAFO trolls, they meet with cartoon dogs, making their response seem silly. Ulyanov's reply set off a chain reaction, and the "pack of Fellas" started posting memes and replies in the comment field, or bonking. Instead of engaging in a lengthy online discussion, the Fellas can set the record straight and dismantle the Russian influence operations using memetic warfare (see Figures 4.35–4.40). During these activities, the Fellas match the dis-and-misinformation with counter-absurd memes or facts (Michaels, 2022; Shultz, 2022; Kirichenko, 2023a)

I mean, your armed forces have been systematically raping, mudering, and deporting Ukrainians on a damned near industrial scale, but sure, this is the "nonsense."

Figure 4.35 NAFO Reaction (a)

How does it feel to be mocked, on such an industrial scale, by a load of Twitter users with cartoon dogs for profile pictures?

Figure 4.36 NAFO Reaction (b)

You know what's nonsense? Getting dragged by cartoon dogs because your country is such a pariah that we're your audience.

Figure 4.37 NAFO Reaction (c)

Yesterday I wrote to an opponent (who misinterpreted my words): "You pronounced this nonsense. Not me." Look at the comments to my tweet below. They return this phrase on every occasion. Many of them are apparently bots, but some seem to be humans. How differentiate between them?

Figure 4.38 Ulyanov's Reply

Ambassador Ulyanov is upset that he is being bullied by cartoon dogs on the internet and is locking his comments to prevent the cartoon dogs from bullying him.

Figure 4.39 NAFO Reaction (d)

Diplomatic update from the Russian Federation, the cartoon dog bullying is intensifying so it is premature to unlock his comments. More on this developing story in the coming days.

Figure 4.40 NAFO Reaction (e)

Ulyanov's engagement in the "shitposting" invited the Fellas pack to rally together (Article 5). They went after him as a group and exposed the apparent foolishness of most Russian disinformation being circulated. Replies and tweets rapidly spread online, and the "nonsense" sentence from Ulyanov's first reply was repeated. This sentence became a slogan for NAFO, and it is possible to buy t-shirts, mugs, and other memorabilia with this on; the proceeds go to Ukraine (Drummond, 2022; Gault, 2022b; Humski, 2022; Michaels, 2022).

Apart from trolling, the NAFO Fellas are also deeply involved in fundraising for the Ukrainian army, using their memes and actions. The group network and individuals are linked to several Ukraine-backed charities and have also been fundraising for signmyrocket.com. They successfully raised funds to name the 2S7Pion "Super Bonker 9000", fully designed with NAFO memes and symbols (Gault, 2022a; Humski, 2022; Taylor, 2022) Operation Raccoon's Revenge, a fundraising event initiated by NAFO, was held to collect money to buy a NAFO water drone for U24 #FleetOfNavalDrones. Community contributions met the required amount by the end of December 2022 (see Figure 4.41) (U24, 2022).

The Usefulness of NAFO

The Russian ability to circulate official "truth" has, over the years, had an impact, as it is difficult to stop these communications if they are not called out immediately. The strength of NAFO is that they are working in an unofficial

subscribers have completed the fundraiser for a naval drone, choosing the eponymous and symbolic name.
Thanks for joining in and puting such a wonderful gift under■■ Christmas tree.
Meanwhile, the #FleetOfNavalDrones fundraiser continues:
u24.gov.ua/navaldrones

UNITED24

0:04 49.5K views

Figure 4.41 U24 Fundraising

context. Therefore, the individual actors in the collective can engage with their targets in a way that states and state actors, internally or externally, cannot. With some care for the language and images used in the memes, the social media company determines the terms and conditions for using the platforms. As long as the NAFO uploads memes, reports tweets to moderators, and calls out propaganda in comment fields, they are not violating any laws and can continue with memetic warfare (Humski, 2022). However, the NAFO Fella community has emerged as the first information operation launched to counter the Russian propaganda machine (Braun, 2022; Shultz, 2022). Some campaigns by the Fellas are reported online and by printed news outlets, and on TV, where the group obtains some celebrity status

Although the NAFO movement is limited to English, it has gained significant development in the Global North. It is, however, debatable how much influence the group has within Russia, where the state controls the flow of online and offline information. Nonetheless, the collective has caused confusion and anger among Russians and their supporters on Twitter. The collective's power and actions have constantly targeted, ridiculed, corrected, and reported Russian propaganda in the comment fields. The Fellas are active in archiving attention and annoyance from those targeted. However, they are often blocked from the targets' websites and accounts online or spend time in "Twitter prison" if their accounts are suspended. The NAFO group appears unstructured, informal, and ad hoc. In the future, there could be a plan to enact private voluntary communities, including NAFO, where it is possible to use the group's flexibility and responsiveness in counter-propaganda (Humski, 2022).

Online Users' Support. Charities and Donations

Numerous fundraising and support groups and networks have been established to assist Ukraine. Fundraising websites, crowdfunding, and charities are conducted online. People worldwide donate money, warm clothing, provide homes for Ukrainian refugees, protest against Russia and its aggressions, and boost morale among Ukrainians (Bernot & Childs, 2022; Dominauskaitė & Svidraitė, 2022). Online platforms and meme culture are practical tools for circulating information about campaigns and crowdsourcing. A specific meme has created a sense of unity within Ukraine, contributing to its resistance and defence and quickly gaining popularity. The meme is linked to Saint Javelin, an image of Madonna Kalashnikov (the Blessed Virgin Mary), which is a modified version of a 2012 print by American artist Chris Shaw. The new version replaces the original AK-47 with an American anti-tank missile (Javelin) (see Figure 4.42). The image celebrates a weapon used in the war and has gained momentum as the visual symbol for the charity Saint Javelin, which operates in support of Ukraine (Haddow, 2016; Antoniuk, 2022; BBC News, 2022; Brzozowski, 2022; Dart, 2022; Debusmann Jr, 2022; KnowYourMeme, 2022;).

As memes become powerful tools in the war, memetic warfare presents a unique opportunity for individuals or charities, such as Saint Javelin, to control the means of media production and collect funds to support Ukraine (Haddow, 2016). Some of these charities' activities have rattled Russian authorities. Saint Javelin, for example, has angered the Russian government, resulting in the artist behind the visual being placed on Russia's official blacklist (Barghouty, 2022; Ministry of Foreign Affairs of the Russian Federation, 2022). The use of memes can have a snowball effect; when online users see the memes, they are reminded of the situation in Ukraine, which can lead to an increase in donations, and users may even be inspired to participate in an event or organise donations themselves (Dominauskaitė & Svidraitė, 2022).

Figure 4.42 Saint Javelin Image

The business side of Saint Javelin has expanded significantly, with a portion of the profits used to support Ukraine. The website features a selection of weapon-carrying saints, branded stickers, accessories, and a donation site (Barghouty, 2022). The Saint Javelin image can be found on many laptops, t-shirts, murals, and tattoos. Additionally, the company behind the charity has turned other significant memes into different gadgets, such as the "Ghost of Kyiv", "Patron, the bomb-sniffing dog", and "NAFO Fella with HIMARS", along with other symbols of Ukrainian war successes (Hogg, 2022).

The funds raised by commerce are directed towards various causes, including mental health charities, purchasing ambulances and anaesthesia devices, and supporting the drone programme's armour purchases. President Zelenskyy launched another crucial initiative, United24, which is the main donation website supporting Ukraine (see Figure 4.43). The website aims to reach out to the online community to gain support and enhance resistance against the Russian aggressor by using international money transfers in an interconnected world. Funds are transferred to the accounts of the National Bank of Ukraine, from where the ministers allocate the money to cover pressing needs to support defence and demining, medical aid and rebuilding Ukraine (Barghouty, 2022; Saint Javelin, 2022; United24, 2022b).

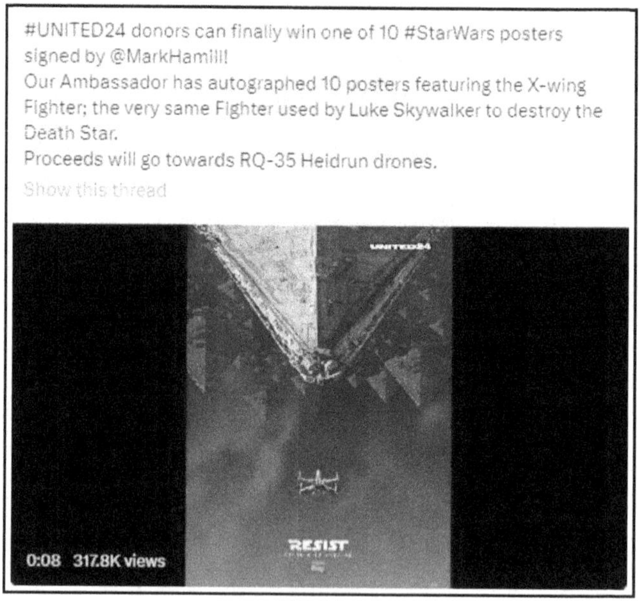

Figure 4.43 U24 Fundraising

There is a high level of transparency regarding the funds donated and their use. Singer and actor Barbra Streisand and Star Wars actor Mark Hamill are ambassadors for the United24 project. Hamill, referring to his role as Luke Skywalker, advocates for more drones to fight off the Russian army (Dugan, 2022; United24, 2022a).

References

Adams, P., 2022. *How Ukraine is winning the social media war.* [Online]
Available at: https://www.bbc.co.uk/news/world-europe-63272202 [Accessed 26 12 2022].
American Institute of Physics, 2022. *Statistical physics rejects theory of "two Ukraines".* [Online] Available at: https://phys.org/news/2022-05-statistical-physics-theory-ukraines.html [Accessed 05 01 2023].
Antoniuk, D., 2022. *Making sense of Ukrainian war memes: From watermelons to Saint Javelin.* [Online] Available at: https://kyivindependent.com/national/making-sense-of-ukrainian-memes-from-watermelons-to-saint-javelin [Accessed 26 12 2022].
Anton, M., 2022. *IT Army of Ukraine to fight on the cyber front via Telegram.* [Online] Available at: https://techthelead.com/it-army-of-ukraine-to-fight-on-the-cyber-front-via-telegram [Accessed 19 09 2022].
Bachman, J., 2022. *U.S. Sends 5,000 SpaceX Starlink internet terminals to Ukraine.* [Online] Available at: https://www.bloomberg.com/news/articles/2022-04-06/u-s-sends-5-000-spacex-starlink-internet-terminals-to-ukraine [Accessed 05 01 2023].

Barghouty, L., 2022. *How the St. Javelin meme raised a million dollars for Ukraine.* [Online] Available at: https://www.washingtonpost.com/world/2022/09/18/ukraine-war-meme-fundraising/ [Accessed 20 09 2022].

BBC News, 2022. *How "Saint Javelin" raised over $1m for Ukraine.* [Online] Available at: https://www.bbc.co.uk/news/world-us-canada-60700906 [Accessed 28 12 2022].

Benebid, M., 2022. *Communication strategies and media influence in the Russia-Ukraine conflict.* [Online] Available at: https://www.policycenter.ma/sites/default/files/2022-04/PB_25-22_Benabid%20EN.pdf [Accessed 12 11 2022].

Bernot, A. & Childs, A., 2022. *Social media in times of war.* [Online] Available at: https://www.lowyinstitute.org/the-interpreter/social-media-times-war [Accessed 16 03 2023].

Bickerton, J., 2022. *Ukraine mocks crying Russian in Crimea with explosions video.* [Online] Available at: https://www.newsweek.com/ukraine-mocks-crying-russians-crimea-explosions-video-saky-air-base-1732826 [Accessed 26 12 2022].

Bishara, H., 2022. *Ukrainians wage a meme war against Russia.* [Online] Available at: https://hyperallergic.com/716738/ukrainians-wage-a-meme-war-against-russia/ [Accessed 01 01 2023].

Braun, S., 2022. *Ukraine's info warriors battling Russian trolls.* [Online] Available at: https://www.dw.com/en/nafo-ukraines-info-warriors-battling-russian-trolls/a-63124443 [Accessed 28 12 2022].

Brzozowski, A., 2022. *Global Europe brief: How Ukraine is winning the meme war.* [Online] Available at: https://www.euractiv.com/section/europe-s-east/news/global-europe-brief-how-ukraine-is-winning-the-meme-war/ [Accessed 19 09 2022].

Burgess, M., 2022. *Ukraine's volunteer "IT Army" is hacking in uncharted territory.* [Online] Available at: https://www.wired.co.uk/article/ukraine-it-army-russia-war-cyberattacks-ddos#:~:text=The%20country%20has%20enlisted% 20thousands, the%20war%20effort%20against%20Russia.&text=Vladimir%20Putin's%20at-tack%20on%20Ukraine, the%20country's%20towns%20and%20cities. [Accessed 27 03 2022].

Cain, G., 2022. *Volodymyr Zelensky on war, technology, and the future of Ukraine.* [Online] Available at: https://www.wired.com/story/volodymyr-zelensky-q-and-a-ukraine-war-technology/?mbid=social_twitter&utm_brand=wired&utm_campaign=falcon&utm_medium=social&utm_social-type=owned&utm_source=twitter&s=09 [Accessed 16 10 2022].

Commander-in-Chief of the Armed Forces of Ukraine, 2022. *@CinC_AFU.* [Online] Available at: https://twitter.com/CinC_AFU [Accessed 30 12 2022].

Dafoe, S., 2022. *"We have proven our strength" – Zelenskyy urges EU to "prove you are with us".* [Online] Available at: https://www.newstalk.com/news/we-have-proven-our-strength-zelenskyy-urges-eu-to-prove-you-are-with-us-1317016 [Accessed 30 12 2022].

Dart, C., 2022. *How a Canadian artist is using the Saint Javelin meme to raise money for Ukraine.* [Online] Available at: https://www.cbc.ca/arts/how-a-canadian-artist-is-using-the-saint-javelin-meme-to-raise-money-for-ukraine-1.6367931 [Accessed 28 12 2022].

Debusmann Jr, B., 2022. *How "Saint Javelin" raised over $1m for Ukraine.* [Online] Available at: https://www.bbc.co.uk/news/world-us-canada-60700906 [Accessed 15 10 2022].

Defence of Ukraine, 2012. *@DefenceU.* [Online] Available at: https://twitter.com/DefenceU [Accessed 04 01 2023].

Defence of Ukraine, 2022a. *@DefenceU.* [Online] Available at: https://twitter.com/ DefenceU/status/1580090899228418048 [Accessed 13 11 2022].

Defence of Ukraine, 2022b. *@DefenceU.* [Online] Available at: https://twitter.com/ DefenceU/status/1587738320477503489 [Accessed 13 11 2022].

Defence of Ukraine, 2022c. *@DefenceU.* [Online] Available at: https://twitter.com/ defenceu/status/1538169555108896776?lang=en [Accessed 13 11 2022].

Defence of Ukraine, 2022d. *@DefenceU.* [Online] Available at: https://twitter.com/ DefenceU/status/1539671779312033793?s=20&t=J0-ddBYutQSVkDcLWzD71g [Accessed 13 11 2022].

Defence of Ukraine, 2022e. *@DefenceU.* [Online] Available at: https://twitter.com/ defenceu [Accessed 13 11 2022].

Defence of Ukraine, 2022f. *@DefenceU.* [Online] Available at: https://twitter.com/ DefenceU/status/1607290106242830339 [Accessed 26 12 2022].

Defence of Ukraine, 2022g. *@DefenceU.* [Online] Available at: https://twitter.com/ defenceu/status/1598283676604895232 [Accessed 26 12 2022].

Defence of Ukraine, 2022h. *@DefenceU.* [Online] Available at: https://twitter.com/ DefenceU/status/1608961253308784642 [Accessed 02 01 2023].

Deighton, K., 2022. *Tripadvisor, Google Maps suspend reviews of some Russian listings.* [Online] Available at: https://www.wsj.com/livecoverage/russia-ukraine-latest-news-2022-03-02/card/tripadvisor-google-maps-suspend-reviews-of-some-russian-listings-vM2no1PgGDmMkL2TSvPZ [Accessed 19 09 2022].

Dominauskaitė, J. & Svidraitė, J., 2022. *The Russo-Ukrainian war prompted people to create memes and here are 35 that show support to Ukraine.* [Online] Available at: https://www.boredpanda.com/ukraine-support-memes/?utm_source= google&utm_ medium=organic&utm_campaign=organic [Accessed 26 12 2022].

Drummond, M., 2022. *Ukraine's internet army of "fellas" are using dog memes to fight Russian propaganda – and they've raised $1m for the army too.* [Online] Available at: https://news.sky.com/story/ukraines-internet-army-of-fellas-are-using-dog-memes-to-fight-russian-propaganda-and-theyve-raised-1m-for-the-army-too-12729625 [Accessed 26 12 2022].

Dugan, E., 2022. *Mark Hamill calls for more drones for Ukraine to fight Russian invasion.* [Online] Available at: https://www.theguardian.com/film/2022/oct/02/ mark-hamill-star-wars-calls-for-more-drones-for-ukraine-to-fight-russian-invasion [Accessed 15 10 2022].

Dzyga's Paw, 2022. *The paw of help to Ukraine.* [Online] Available at: https://dzygas-paw.com/ [Accessed 21 12 2022].

Ellwood, D., 2022. *Narratives, propaganda & "smart" power in the Ukraine conflict, Part 2: Inventing a global presence.* [Online] Available at: https://uscpublicdiplo-macy.org/blog/pd-wartime-narratives-propaganda-smart-power-ukraine-conflict-part-2-inventing-global [Accessed 13 11 2022].

Fedorov, M., 2022. *Twitter.* [Online] Available at: https://twitter.com/FedorovMykhailo/ status/1497642156076511233?ref_src=twsrc%5Etfw%7Ctwcamp%5Etweetembe d%7Ctwterm%5E1497642156076511233%7Ctwgr%5Ecec632913f1b58dc65ca9 6eabc3e4bf32c70460%7Ctwcon%5Es1_&ref_url=https%3A%2F%2Ftechthelead. com%2Fit-army-of-ukra [Accessed 19 09 2022].

Fiala, O., 2022. Resilience and Resistance in Ukraine. *Small Wars Journal,* 31, p. 12.

Gault, M., 2022a. *NAFO memesters paid Ukraine to paint their memes on a tank.* [Online] Available at: https://www.vice.com/en/article/epzp7n/nafo-memesters-paid-ukraine-to-paint-their-memes-on-a-tank [Accessed 26 12 2022].

Gault, M., 2022b. *Shitposting Shiba Inu accounts chased a Russian diplomat offline.* [Online] Available at: https://www.vice.com/en/article/y3pd5y/shitposting-shiba-inu-accounts-chased-a-russian-diplomat-offline [Accessed 27 12 2022].

General Staff of the Armed Forces of Ukraine, 2015. *@GeneralStaffUA.* [Online] Available at: https://twitter.com/GeneralStaffUA [Accessed 29 03 2023].

Griffin, A., 2022. *Hundreds of thousands join Ukraine "IT Army" to fight cyberwar with Russia.* [Online] Available at: https://www.independent.co.uk/tech/russia-ukraine-cyber-attack-it-army-telegram-b2024927.html [Accessed 19 09 2022].

Gronholt-Pedersen, J., 2022. *Keyboard army using restaurant reviews to take on Russian state media.* [Online] Available at: https://www.reuters.com/world/europe/keyboard-army-using-restaurant-reviews-take-russian-state-media-2022-03-02/ [Accessed 10 10 2022].

Haddow, D., 2016. *Meme warfare: How the power of mass replication has poisoned the US election.* [Online] Available at: https://www.theguardian.com/us-news/2016/nov/04/ political-memes-2016-election-hillary-clinton-donald-trump [Accessed 25 2 2022].

Haji, R., McKeown, S., & Ferguson, N., 2016. Social Identity and Peace Psychology. In: *Understanding Peace and Conflict Through Social Identity Theory.* Cham: Springer Nature, pp. XV–XX.

Herszenhorn, D.M. & McLeary, P., 2022. *Ukraine's "iron general" is a hero, but he's no star.* [Online] Available at: https://www.politico.com/news/2022/04/08/ukraines-iron-general-zaluzhnyy-0002390 [Accessed 05 01 2022].

Hogg, M.A., 2016. Social Identity Theory. In: *Understanding Peace and Conflict Through Social Identity Theory.* Cham: Springer Natural, pp. 3–18.

Hogg, R., 2022. *How the founder of the Saint Javelin charity brand worn by Zelenskyy plans to help rebuild Ukraine.* [Online] Available at: https://www.businessinsider.com/ saint-javelin-founder-worn-by-zelensky-plans-help-rebuild-ukraine-2022–5?r=US&IR=T [Accessed 15 10 2022].

Holt, T.J. & Bossler, A.M., 2016. *Cybercrime in progress. Theory and Prevention of Technology-enabled Offenses.* London: Routledge.

Humski, J.C., 2022. *Can the West create a NAFO that's built to last beyond Ukraine?* [Online] Available at: https://thehill.com/opinion/national-security/3782298-can-the-west-create-a-nafo-thats-built-to-last-beyond-ukraine/ [Accessed 27 12 2022].

IMDb, 1988. *Die Hard.* [Online] Available at: https://www.imdb.com/title/tt0095016/ [Accessed 26 12 2022].

Jordan, J. & Taylor, P.A., 2004. *Hacktivism and cyberwards. Rebels with a cause?* Abingdon: Routledge.

Kirichenko, D., 2023a. *Decentralisation is NAFO's greatest strength.* [Online] Available at: https://emerging-europe.com/voices/why-decentralisation-is-nafos-greatest-strength/ [Accessed 11 03 2023].

Kirichenko, D., 2023b. *NAFO's Fellas must evolve.* [Online] Available at: https://cepa.org/article/nafos-fellas-must-evolve/ [Accessed 10 03 2023].

KnowYourMeme, 2022. *St. Javelin/Saint Javelin.* [Online] Available at: https://knowyourmeme.com/memes/st-javelin-saint-javelin [Accessed 19 09 2022].

Lavorgna, A., 2020. *Cybercrimes. Critical issues in a global context.* London: Macmillian.

Maskaliūnaitė, A., 2021. Exploring Resistance Operating Concept. Promises and Pitfalls of (Violent) Underground Resistance. *Journal on Baltic Security,* 7(1), pp. 27–38.

McInnis, K., Jones, S.G., & Harding, E., 2022. *NAFO and winning the information war: Lessons learned from Ukraine.* [Online] Available at: https://www.csis.org/analysis/ nafo-and-winning-information-war-lessons-learned-ukraine [Accessed 27 12 2022].

McNair, B., 2011. *An Introduction to Political Communication.* 5th ed. London: Routledge.

Michaels, D., 2022. *Ukraine's internet army of "NAFO Fellas" fights Russian trolls and rewards donors with dogs.* [Online] Available at: https://www.wsj.com/ articles/ ukraines-internet-army-of-nafo-fellas-fights-russian-trolls-and-rewards-donors-with-dogs-11664271002 [Accessed 11 03 2023].

Mikhail_Ulyanov, 2022. *@Amb_Ulyanov.* [Online] Available at: https://twitter.com/ amb_ulyanov/status/1538562863199141889?lang=en [Accessed 27 12 2022].

Milmo, D., 2022. *Amateur hackers warned against joining Ukraine's "IT army".* [Online] Available at: https://www.theguardian.com/world/2022/mar/18/amateur-hackers-warned-against-joining-ukraines-it-army [Accessed 27 03 2022].

Ministry of Digital Transformation of Ukraine, 2022. *Ministry of Digital Transformation: IT army blocks Russian sites in a few minutes – the main victories of Ukraine on the cyber front.* [Online] Available at: https://www.kmu.gov.ua/en/news/mincifri-it-armiya-blokuye-rosijski-sajti-za-dekilka-hvilin-golovni-peremogi-ukrayini-na-kiberfronti [Accessed 05 01 2022].

Ministry of Foreign Affairs of the Russian Federation, 2022. *Canadian citizens under personal sanctions, including a ban on entry into the Russian Federation.* [Online] Available at: https://www.mid.ru/ru/detail-material-page/1811224/ [Accessed 19 09 2022].

Molchanov, M.A., 2016. Russia as Ukraine's "Other": Identity and geopolitics. In: *Ukraine and Russia. People, Politics, Propaganda and Perspectives.* Bristol: E-International Relations Publishing, pp. 195–210.

Mortensen, M. & Neumayer, C., 2021. The Playful Politics of Memes. *Information, Communication and Society*, 24(16), pp. 2367–2377.

Munk, T., 2022. *The Rise of Politically Motivated Cyber Attacks.* London: Routledge.

Munk, T. & Ahmad, J., 2022. "I Need Ammunition, Not a Ride": The Ukrainian cyber war. *Comunicação e Sociedade*, 42, pp. 221–241.

Nicholson, K., 2022. *Hacker group Anonymous claim they interrupted Russian TV with harrowing footage of Ukraine.* [Online] Available at: https://www.huffing-tonpost.co.uk/entry/anonymous-hacker-russian-state-tv-ukrainian-bombing_uk_6225f375e4b04a0545d98d62 [Accessed 19 09 2022].

North Atlantic Fella Organization, 2022. *@Official_NAFO.* [Online] Available at: https://twitter.com/Official_NAFO/status/ 1577269391997050880? ref_src=twsrc %5Etfw%7Ctwcamp%5Etweetembed%7Ctwterm%5E1577274895095906305% 7Ctwgr%5E60338d1736a091a06999d94fc562e0c9cc56eff2%7Ctwcon%5Es2_& ref_url=https%3A%2F%2Fnews.sky.com%2Fstory%2Fukraines-int [Accessed 27 12 2022].

OECD, 2022. *Disinformation and Russia's war of aggression against Ukraine.* [Online] Available at: https://www.oecd.org/ukraine-hub/policy-responses/disinformation-and-russia-s-war-of-aggression-against-ukraine-37186bde/ [Accessed 03 01 2023].

Oleksii Reznikov, 2015. *@oleksiireznikov.* [Online] Available at: https://twitter.com/ oleksiireznikov [Accessed 04 01 2023].

Oleksii_Reznikov, 2022. *@oleksiireznikov.* [Online] Available at: https://twitter.com/ oleksiireznikov/status/1609160243597934593 [Accessed 31 12 2022].

Onuch, O. & Hale, H. E., 2022. *The Zelensky Effect.* London: Hurst.

Pantic, V., 2022. *How to support Ukraine with community engagement.* [Online] Available at: https://www.citizenlab.co/blog/civic-engagement/how-to-support-ukraine-with-community-engagement/ [Accessed 25 12 2022].

Parker, J. & Kremling, A.M.S., 2017. *Cyberspace, Cybersecurity, and Cybercrime.* Thousand Oaks: SagePublications.

Pitrelli, M., 2022. *Hacktivist group Anonymous is using six top techniques to "embarrass" Russia.* [Online] Available at: Hacktivist group Anonymous is using six top techniques to "embarrass" Russia [Accessed 30 12 2022].

Powell, M., 2022. *Now I have HIMARS, ho ho ho! Ukraine compares its fight against Putin to Die Hard in social media video – and insists the "scrappy underdog" WILL win.* [Online] Available at: https://www.dailymail.co.uk/news/article-11574361/Ukraine-compares-fight-against-Putin-Die-Hard-social-media-video.html [Accessed 26 12 2022].

Rating Group, 2022. *The eight national poll: Ukraine during the war.* [Online] Available at: https://ratinggroup.ua/en/research/ukraine/ vosmoy_obschenacionalnyy_opros_ukraina_v_usloviyah_voyny_6_aprelya_2022.html [Accessed 25 12 2022].

Riabchuk, M., 2015. "Two Ukraines" Reconsidered: The end of Ukrainian ambivalence? *Studeis in Ethnicity and Nationalism,* 15(1), pp. 138–156.

Ross, A.S. & Rivers, D.J., 2017. Digital Cultures of Political Participation: Internet memes and thediscursive delegitimization of the 2016 U.S presidential candidates. *Discourse, Context & Media,* 16, pp. 1–11 (https://www.sciencedirect.com/science/article/abs/pii/ S2211695816301684?via%3Dihub).

Sabbagh, D., 2022. *Ukrainian attack on Russian airbase sends message to Moscow and beyond.* [Online] Available at: https://www.theguardian.com/world/2022/aug/11/ukraine-attack-russian-airbase-saky-crimea-send-message-moscow-propaganda-victory [Accessed 29 03 2023].

Sabbagh, D. & Lock, S., 2022. *Russian warplanes destroyed in Crimea airbase attack, satellite images show.* [Online] Available at: https://www.theguardian.com/world/2022/aug/11/russian-warplanes-destroyed-in-crimea-saky-airbase-attack-satellite-images-show [Accessed 26 12 2022].

Saint Javelin, 2022. *Saint Javelin.* [Online] Available at: https://www.saintjavelin.com/ [Accessed 21 12 2022].

Screti, F., 2013. Defending Joy Against the Popular Revolution: Legitimation and delegitimation through songs. *Critical Discourse Studies,* 10(2), pp. 205–222.

Serrano, J., 2022. *Ukraine creates IT army of volunteer hackers and orders cyber attacks on Russian websites.* [Online] Available at: https://gizmodo.com/ukraine-it-volunteer-hacker-army-response-to-russian-in-1848600395 [Accessed 19 09 2022].

Shultz, B., 2022. *Meme warfare: What western governments can learn from the NAFO Alliance.* [Online] Available at: https://intpolicydigest.org/meme-warfare-what-western-governments-can-learn-from-the-nafo-alliance/ [Accessed 31 12 2022].

Shuster, S. & Bergengruen, V., 2022. *Inside the Ukrainian counterstrike that turned the tide of the war.* [Online] Available at: https://time.com/6216213/ukraine-military-valeriy-zaluzhny/ [Accessed 04 01 2023].

Sirikupt, C., 2022. *What's so funny about a Russian invasion?* [Online] Available at: https://www.washingtonpost.com/politics/2022/04/07/ukraine-russia-memes-satire-humor/ [Accessed 07 04 2023].

Smith, A., 2022. *Google Maps suspends reviews as Russian landmarks flooded with photos of captured soldiers and news clips.* [Online] Available at: https://www.independent.co.uk/tech/google-maps-russian-landmarks-photos-reviews-b2027638.html [Accessed 19 09 2022].

Snowden, C., 2022. *Guns, tanks and Twitter: How Russia and Ukraine are using social media as the war drags on.* [Online] Available at: https://theconversation.com/guns-tanks-and-twitter-how-russia-and-ukraine-are-using-social-media-as-the-war-drags-on-180131 [Accessed 13 11 2022].

Srivastava, M., Miller, C., & Olearchyk, R., 2022. *"Trolling helps show the king has no clothes": How Ukraine's army conquered Twitter.* [Online] Available at: https://www.ft.com/content/b07224e1-414c-4fbd-8e2f-cfda052f7bb2 [Accessed 13 11 2022].

Steffek, J., 2003. The Legitimation of International Governance: A discourse approach. *European Journal of International Relations,* 9(2), pp. 249–275.

Taylor, A., 2022. *With NAFO, Ukraine turns the trolls on Russia.* [Online] Available at: https://www.washingtonpost.com/world/2022/09/01/nafo-ukraine-russia/ [Accessed 01 09 2022].

Teh, C., 2022. *Hackers changed the call sign of a Putin-linked superyacht to "FCKPTN" and set the ship's destination as "hell".* [Online] Available at: https://www.insider.com/hackers-change-call-sign-of-putin-linked-superyacht-to-fckptn-2022-3 [Accessed 26 12 2022].

U24, 2022. *@U24_gov.ua.* [Online] Available at: https://twitter.com/U24_gov_ua/ status/1608499260143853569 [Accessed 29 12 2022].

Ukraine, 2016. *@Ukraine.* [Online] Available at: https://twitter.com/ukraine [Accessed 13 11 2022].

Ukraine Memes for NATO Teens, 2022. *@LivFaustDieJung.* [Online] Available at: https://twitter.com/LivFaustDieJung [Accessed 31 12 2022].

Ukraine/Україна, 2022. *@Ukraine.* [Online] Available at: https://twitter.com/Ukraine [Accessed 26 12 2022].

Ukrainer, 2022. *Who are the NAFO Fellas? The army of cartoon dogs fighting Russian propaganda.* [Online] Available at: https://ukrainer.net/nafo-fellas/ [Accessed 27 12 2022].

Ukrainian Meme Force, 2022. *@uamemesforce.* [Online] Available at: https://mobile.twitter.com/uamemesforces [Accessed 01 01 2023].

United24, 2022a. *Five months of United24 activity.* [Online] Available at: https://u24.gov.ua/news/five_months_of_activity [Accessed 16 10 2022].

United24, 2022b. *The initiative of the President of Ukraine.* [Online] Available at: https://u24.gov.ua/ [Accessed 21 12 2022].

Yar, M. & Steinmetz, K.F., 2019. *Cybercrime and Society.* 3rd ed. London: Sage.

Yekelchyk, S., 2020. *Ukraine. What Everyone Needs to Know.* 2nd ed. Oxford: Oxford University Press.

Zaliznyak, Y., 2022. *Ukraine's volunteer "IT army" responds to Russian hackers, minister says.* [Online] Available at: https://abcnews.go.com/International/ukraines-volunteer-army-responds-russian-hackers-minister/story?id=88651955 [Accessed 26 12 2022].

Zanin, M. & Martínez, J.H., 2022. Analyzing International Events Through the Lens of Statistical Physics: The case of Ukraine. *Chaos: An Interdisciplinary Journal of Nonlinear Science,* 32(5), pp. 1–9.

Zarembo, K., 2022. *Civic activism against geopolitics: The case of Ukraine.* [Online] Available at: https://carnegieeurope.eu/2022/11/30/civic-activism-against-geopolitics-case-of-ukraine-pub-88485 [Accessed 25 12 2022].

5 Memetic Warfare

Communication as a Weapon

Tine Munk

Behaviours and Expressions Matter

Language and symbols are used unambiguously to define and communicate category membership through markers or shared symbols and implicit clues. Once categorisation and self-categorisation are activated, several behaviours, including stereotyping, consensus seeking, normative reasoning, and collective actions, lead to practices that solidify and institutionalise cultures and subcultures during the war as the country has a common enemy (Keblusek et al., 2017, p. 3). Political language and communication encompass rhetoric and paralinguistics, such as body language, or political acts, such as war, resilience, and protest (Graber, 1981; McNair, 2011, p. 3). Language is not necessarily limited to verbal expressions; it can also be linked to body language, performance, and behaviours. When analysing memes, it becomes clear that the body language of key actors included in the memes is essential. For example, specific cultural and political contexts, politicians' bodies, where they are, what they wear, how they sit, how they walk, talk, etc. (Diehl, 2021). This is important for portraying certain political persons in the memes to show what they should symbolise – are they inclusive or exclusive figures, and how can this be implemented to generate a feeling of togetherness of the in-group and distance the out-group?

Impressions and Behaviours

The impressions, expressions, and behaviours of the two wartime presidents send a signal to the audience and serve as inspiration behind memes in memetic warfare. Memesters analyse what these men say and do and how the images are staged to create new and different memes with catchy text. With the participatory nature of social media and virtual space, online images, creativity, and innovations have become the core elements of memetic war. Memes have become a social phenomenon that engages several online users in creating, altering, and spreading posts. The memes circulated are not only for fun and mocking but also to draw attention to poignant cultural

DOI: 10.4324/9781003432630-5

and political situations or events (Barnes et al., 2021). Memes can also be used to express support for a particular cause or to criticise the actions of an opposing group. As memes spread rapidly through social media platforms, they can quickly reach a broad audience and become a powerful tool for shaping public opinion. This makes it increasingly difficult for governments and other actors to control the narrative of a conflict or to shape public opinion in their favour. As a result, memetic warfare has become an essential

Figure 5.1 President Zelenskyy and President Putin (a)

component of modern warfare, and understanding how memes are used to shape public opinion is crucial for anyone studying or participating in the conflict (Seib, 2021, p. 95).

The media and social media should not be underestimated in the context of memetic warfare. How the two leaders approach the online environment can be crucial for winning the war. Memes, which often include an image and short implied text, require the audience to decode the meaning quickly (Grundlingh, 2018, p. 154). Therefore, it is essential that the link between the image and what the meme is trying to communicate is easy to understand. The images included are also important, as the audience should be able to understand them without much introduction. However, memes can remain relevant because memesters constantly repossess them. Memesters can select images to pursue a particular outcome, and these selections are not based on objective criteria but rather are orchestrated to achieve a specific goal. Memes featuring President Putin may not have much effect in Russia, where the state controls the online space. However, outside Russia, they serve as a constant reminder of the differences between the two presidents (see Figures 5.1–5.4). The Ukrainian/ Ukraine-supportive memesters can win the memetic war through their creativity and precise character assassination. The division

Figure 5.2 President Zelenskyy and President Putin (b)

BRAVERY
BRUTALITY

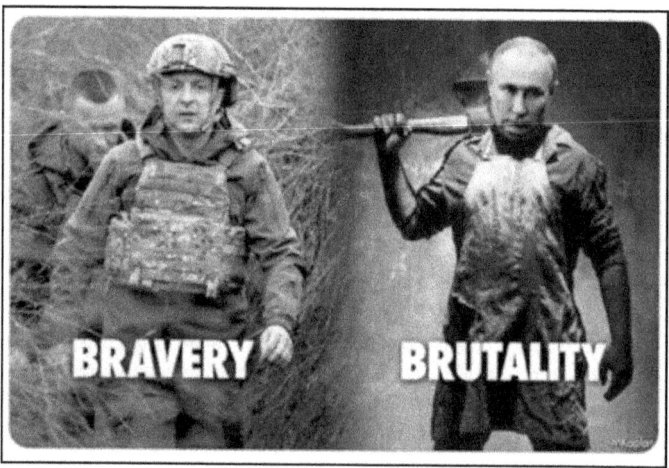

Figure 5.3 President Zelenskyy and President Putin (c)

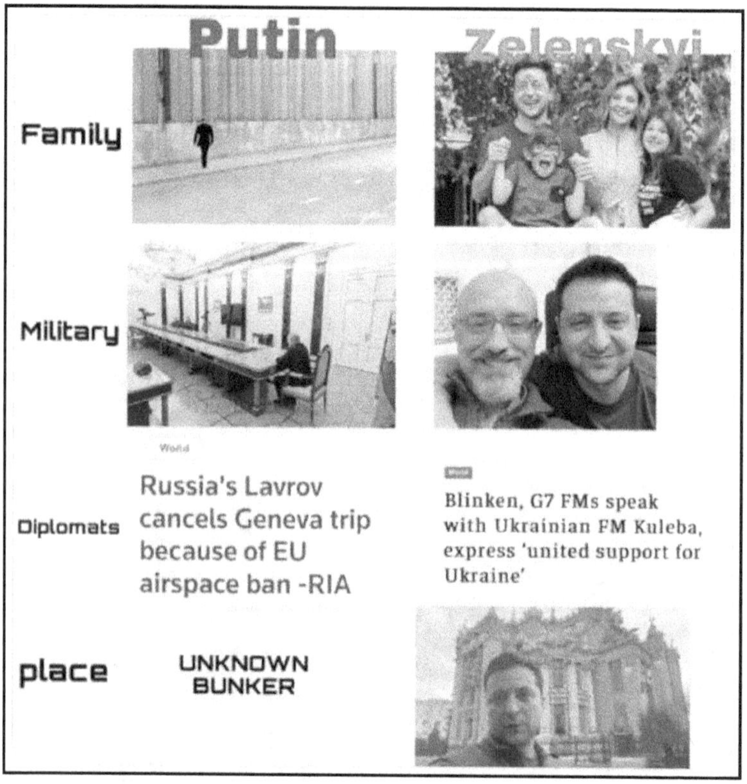
Putin
Zelenskyi
Family
Military
Diplomats
Russia's Lavrov cancels Geneva trip because of EU airspace ban -RIA
Blinken, G7 FMs speak with Ukrainian FM Kuleba, express 'united support for Ukraine'
place
UNKNOWN BUNKER

Figure 5.4 President Zelenskyy and President Putin (d)

between in-groups and out-groups is evident in how the presidents are portrayed in memes, with each representing the core elements of their states, values, and desires.

President Zelenskyy

Undoubtedly, President Zelenskyy won the media war through his masterful communication skills. He reaches out to his audience in small videos where he responds defiantly to the invasion. His speeches and video recordings provide a vision of an ordinary citizen and a solid and caring leader, essential to keeping the population calm and building resistance. President Zelenskyy has adopted a practical wartime image by wearing t-shirts and jumpers with symbols of Ukraine, such as the Armed Forces, the Ukrainian flag, or the Trident. These symbols and speeches are repeatedly incorporated into memetic warfare and have become a fashion symbol (Genauer, 2022; Munk & Ahmad, 2022, p. 229; Onuch & Hale, 2022; Segal, 2022). President Zelenskyy does not wear a suit out of principle and has not done so since the start of the war. Neither does he wear standard military uniform, decorations, or insignias, just as his government officials do not. The casual military style signals that no one is above the people, and all are in the trenches together (Genauer, 2022; Segal, 2022).

President Putin

The memes created by Ukrainian citizens and supporters highlight the differences between the two leaders, portraying President Zelenskyy as dignified and warm while depicting President Putin as distant and withdrawn. Images of Putin in the mass media often portray him as pale, cold, withdrawn, aggressive, erratic, and spiteful, in contrast to Zelenskyy's more relaxed and approachable appearance. Putin is often shown sitting far from others in a large room or behind a long table, whereas Zelenskyy is portrayed as standing close to the people, creating a sense of togetherness (CBS News, 2022; Guardian, 2022a; Holmes, 2022; Mulvey, 2022; Munk & Ahmad, 2022, p. 229; Sauer, 2022a; Saul, 2022; Smith, 2022). The Kremlin attempted to soften Putin's image by arranging a "unity" rally in Moscow, but his appearance was still perceived as less warm and energetic than that of Zelenskyy. Putin has mostly only appeared with selected groups of people, such as state leaders or mothers of Russian soldiers, but these images have not had the intended calming effect. For example, in November 2022, Putin appeared to talk to a group of mothers of Russian soldiers serving in the army. These women were hand-picked to avoid crucial questions being posed to the President, but the him did not have the intended effect (Nechepurenko, 2022; Roth & Sauer, 2022). In the world of memes, a positive event or PR opportunity can quickly turn into its opposite. Memesters can take an image or situation that was meant to be positive and turn it into a mockery within seconds. For instance, within

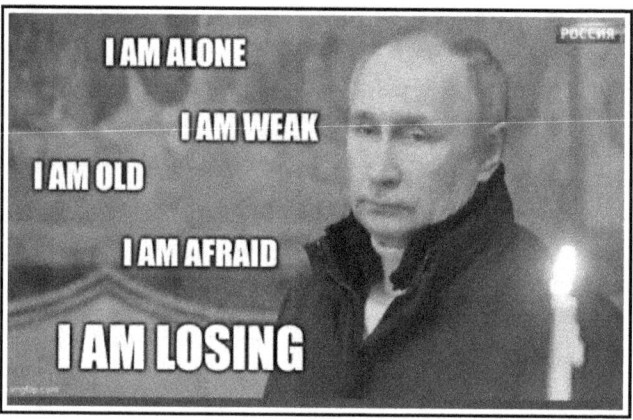

Figure 5.5 Memes of President Putin's Church Visit

minutes of the official picture of President Putin standing alone in a Russian Orthodox Church on Christmas Day (according to the Georgian calendar), which merely highlighted his isolation, memesters had already made the picture available on Twitter for their purposes (Reuters, 2023). Within hours of the image's publication, several memes and videos featuring Putin with various captions began circulating (see Figure 5.5).

Wartime Presidents

President Putin's New Year speech has been significantly mocked on social media. In contrast to previous years, he did not deliver his speech from the Kremlin but instead was surrounded by soldiers in an attempt to create the impression of being closely involved in the war. However, photos from the event were analysed and appeared to show actors, apparatchiks, and functionaries instead of actual soldiers. These individuals had previously acted as extras in other propaganda videos and events. The propaganda event backfired and failed to improve the image of the withdrawn and isolated wartime President (see Figures 5.6, 5.7) (Kika, 2022; Kestenholz, 2023). The Ukrainian Memes Forces created a meme within a short time span showing President Putin behind bars with the fake soldiers behind him, with the caption "war criminals should be in prison" (see Figure 5.10) (Ukrainian Memes Forces, 2022).

Ukraine's President Zelenskyy's New Year speech stands in sharp contrast. He delivers a passionate speech that acknowledges the country's defence efforts, highlights its successes, and addresses the difficult moments of the war (President of Ukraine, 2022; Kyiv Independent, 2023). In his speech, President Zelenskyy stands alone on the streets of Kyiv in the dark. However, President Putin's visit to the base for his New Year speech appears to have been enacted to mirror President Zelenskyy's visit to Kherson after the city's liberation during the autumn 2022 counteroffensive. The liberation of Kherson was a significant setback for Russia, which was forced to retreat to the other side of the Dnipro River (BBC News, 2022; Kyiv Independent, 2022b).

In Ukraine, positive and morale-boosting memes and communications are constantly created to support President Zelenskyy as a statesman and wartime president. One example is President Zelenskyy's unannounced visit to Bakhmut, where Ukrainian and Russian forces fought a long and devastating

Figure 5.6 Putin with Fake Audience

Figure 5.7 Putin with Fake Soldiers

battle known as the "meat grinder". In spite of the ongoing fighting in the area and the security risk involved, President Zelenskyy met with the troops and personally handed out awards to soldiers. This visit was also symbolic as President Zelenskyy left Bakhmut to fly to the USA to speak to the joint Congress in Washington DC (Bachega & Greenall, 2022; Stein et al., 2022). These events have inspired several memes highlighting the contrast between the two presidents' warm and cold expressions and behaviours.

Non-visual Language and Memes

Zelenskyy's Speeches and One-liners

The speeches made by Ukrainian President Zelenskyy have been instrumental since the beginning of the war, with his visual images and verbal communications going viral worldwide. This use of speeches online began with his front-facing video, where he and his cabinet ministers stood in defiance on Kyiv's streets, ready for combat (Knibbs, 2022; Munk & Ahmad, 2022). Commentators expected Ukraine's military and government to collapse, but the government, army, and civilians fought back in the days and months following the 2022 invasion. President Zelenskyy and his cabinet stayed in Kyiv, and he accepted a new and essential role as the face and voice of Ukraine. Through nightly speeches, he demonstrated the Ukrainian people's resistance and defiance (Genauer, 2022; Zelensky, 2022, p. 39).

In his speech on 24 February 2022, Zelenskyy tried to assure the population by talking directly to the audience:

> I will talk to you again soon. Do not panic. We are strong. We are ready for everything. We will defeat anyone. Because we are Ukraine. (Zelensky, 2022, p. 57)

During his speeches, President Zelenskyy cemented Ukraine's independence as a dignified and united country, countering Russia's justification for invading the country. These speeches, shared on social media, gave him a direct and unfiltered route to ordinary citizens, creating a personal connection (Onuch & Hale, 2022, pp. 10–11). Many of his speeches have been turned into memes with changing visuals (see Figures 5.12–5.14). The ethos of these speeches and online posts and the communicator's credibility is crucial. The targeted audience amplifies memes through the emotional anguish of the war and the country's gains and losses. However, these communications share a common ground, as the in-group members, including the communicator and the audience, have shared beliefs, values, and interests (Lipschultz, 2022, p. 94).

Mass media often repeat snippets of speeches, while online communications, such as memes and videos, spread the key message widely to a broader audience (Seib, 2021, p. 161). One example is the quote "I need ammunition, not a ride", which emerged when President Zelenskyy refused an offer from the USA to evacuate from Kyiv during the early days of the war (Braithwaite, 2022; Munk & Ahmad, 2022; Prothero, 2022). This citation has been constantly reused by mass media and shared on social media, where it transformed into the core text of several memes (see Figure 5.8) (Munk & Ahmad, 2022; Onuch & Hale, 2022, p. 9; Wood, 2022).

The same happened with the "nobody is going to break us, we're strong, we are Ukrainians" speech to the European Parliament on 1 March 2022. It was aimed at foreign leaders and institutions in order to gather support, money, and weapons for Ukraine. This speech has also been widely shared on social media and turned into several memes (Genauer, 2022; Wood, 2022; Sky News, 2022; Zelensky, 2022). These memes reinforce President Zelenskyy's image as a strong and committed leader dedicated to defending Ukraine against Russian aggression. By using humour and popular culture references, the memes help to make his message more accessible to a broader audience.

Zelenskyy's ability to capture the world's attention and make sense of a complex situation has been key to his success. He has been able to weave national stories into a narrative that highlights the struggle in Ukraine as

Figure 5.8 President Zelenskyy's Reply to the USA

not only a Ukrainian issue but one that affects the security of the entire free democratic world. In addition, his speeches have emphasised the security issues related to the fight against Russian aggression, which threatens stability and freedom in the Western world (Genauer, 2022). Zelenskyy's speeches also focus on emotions and the security of future generations, not just in Ukraine but also in other countries such as the United States. Many of these speeches contain memorable one-liners that stay in people's minds, which are often repeated in memes and used as part of information warfare and psychological operations to influence behaviour or spread and counter dis-and-misinformation (see Figure 5.9) (Ascott, 2020; President of Ukraine, 2022; Stupples, 2015).

President Zelenskyy's "Without You" Speech

Politicians worldwide have become adept at using short and precise communications to disseminate information that can be easily incorporated into media accounts and circulated quickly and efficiently to a large audience (Schill, 2012, p. 120). Online communications, using both text and visual images,

FREEDOM MUST BE ARMED BETTER THAN TYRANNY

Ukranian President, Volodymyr Zelenskyy

Figure 5.9 Meme Based on a Speech by Zelenskyy

have made it clear that mastering the online environment can be as powerful as traditional war tools, such as bullets, missiles, and tanks. Furthermore, various online communications have a longer shelf life than newspapers, magazines, and TV broadcasts. The visual images and text are reused numerous times in online memes.

President Zelenskyy's powerful "Without You" speech, directed at the Russian people at the beginning of the Autumn counteroffensive in 2022, served as a warning to Moscow while the Ukrainian army began the liberation of occupied areas in the eastern part of Ukraine (Haq et al., 2022; Sussex, 2022).

Key speeches by President Zelenskyy and other prominent figures become part of meme communication. These speeches contribute to the narrative that they are all part of the resistance against Russia and will not surrender, no matter how much hardship they endure due to Russian aggression. These speeches communicate a message of resistance and demonstrate that Ukraine is a united nation that stands together in the face of Russian aggression (see Chapter 2). Parts of the "Without You" speech have become a symbol of resistance. This is a mantra for Ukrainians, who endure appalling

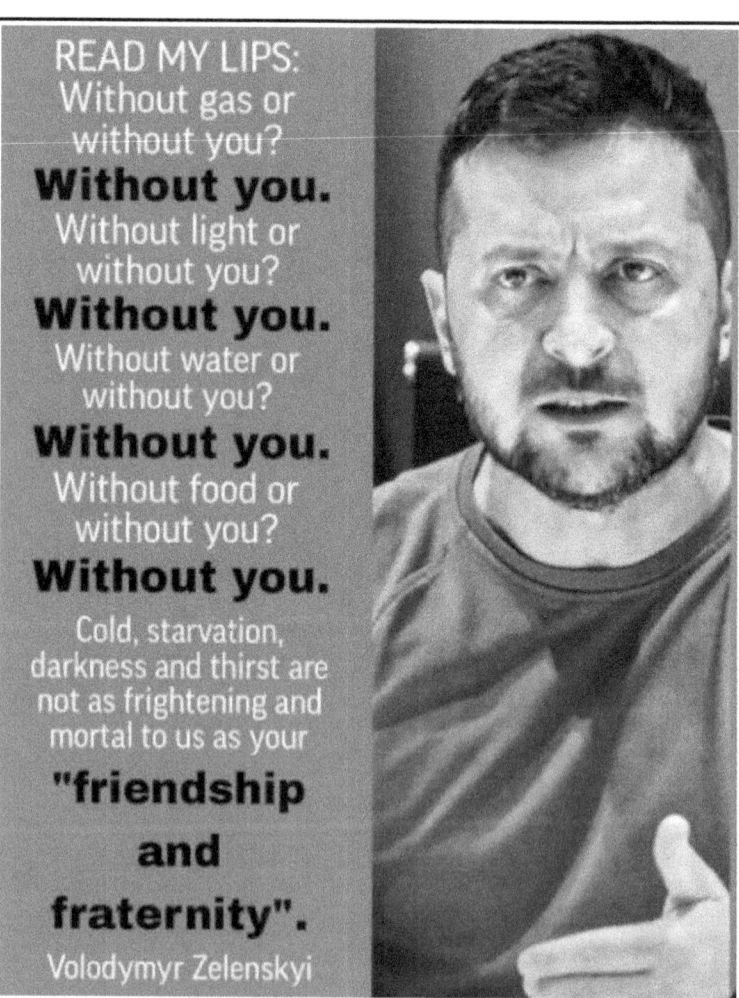

READ MY LIPS:
Without gas or
without you?
Without you.
Without light or
without you?
Without you.
Without water or
without you?
Without you.
Without food or
without you?
Without you.
Cold, starvation,
darkness and thirst are
not as frightening and
mortal to us as your
"friendship
and
fraternity".
Volodymyr Zelenskyi

Figure 5.10 Zelenskyy's Reaction to Intense Russian Shelling of Civilians

life conditions due to the constant shelling and associated power blackouts (Borger, 2022; Ellyatt, 2022). For example, after the liberation of Kherson, one citizen said:

> We are happy to have no electricity and no water, if that's the price of being without the Russians. Can you imagine, there are no more Russians with white armbands on my street! (Mendel, 2022)

The same argument has been forwarded by Kyiv citizens who prefer to be without power compared to living with the Russian occupiers (see Figures 5.11, 5.12) (Leicester et al., 2022).

Memes, Codes and Symbols

Symbols and codes are important in digital culture, where images are crucial for communication and political actions. Symbols create meaning through their visual impact, and their attribution comes from social interactions and the value they hold (Athique, 2015, p. 97). Communication is a dynamic process that evolves over time. Symbols are deeply rooted in this process, enabling the decoding of messages and their context, particularly in memes that rely on short texts and images. However, using memes and symbols together requires online users to have a shared understanding of their meaning

Figure 5.11 Reference to the "Without You" Speech, Winter 2022

They wanted to freeze us and throw us into darkness. We survived!
Today is the first day of spring.
Life, light, love defeat death.
Ukraine will win.

Figure 5.12 Reference to the "Without You" Speech, Spring 2023

to decode their messages. Cohesive groups build transmitted collective identities through creating, receiving, sharing, and repurposing memes in new ways (Mortensen & Neumayer, 2021, p. 2370; Floyd et al., 2022, p. 2).

National Symbols

Language and symbols have taken a central position in the resistance against Russia. The Ukrainian language is crucial to the feeling of togetherness and civic engagement in the resistance. It serves as an identity-building block and is deeply integrated into the sense of being Ukrainian, especially since the full-scale invasion in 2022 (Afanasiev et al., 2022). Nations become visible through different symbols, which are progressively included on numerous levels in Ukraine to enforce a sense of belonging and visualise what is at stake – the state's sovereignty and self-determination for citizens about their life and future. Flags, anthems, and emblems unite society against the common enemy, and all are visible signs of belonging to the in-group (Elgenius, 2011, p. 2). National symbols are not merely images of a state; they represent a larger context in which emotions are attached to shared knowledge, values, history, and memories, all of which are challenged by the Russian invasion (Butz, 2009, p. 780). These national symbols are significant for intergroup relations, where they are perceived positively by the in-group and negatively by the out-group, leading to negative emotions (Muldoon et al., 2020, p. 265). Visuals are a part of everyday life, and viewers tend to believe and process images more than written text and symbols (Schill, 2012, p. 121). Moreover, visual images transmit particular messages more clearly than textual or verbal communication (Schill, 2012, p. 121; Farkas & Bene, 2021, p. 120).

"Slava Ukraini" and "Heroyam Slava"

After the full-scale invasion of Ukraine, the salutations "Slava Ukraini!" [Glory to Ukraine!] and "Heroyam slava!" [Glory to Heroes!] have gained global attention and are often used in tweets and memes to express support for Ukrainians. These salutations are rooted in the Ukrainian national liberation fight from 1918–1922. They gained new significance during the Euromaidan revolution in 2013–2014, where they were used to promote a liberal and pro-European political agenda and break free from past colonial and imperial suppression (Chraibi, 2016; Ukraine World, 2021). The salutations serve as reminders of Ukraine's fight for freedom, democracy, and sovereignty and its role as the first-line defender of democracy and Western values. Today, it is possible to buy clothing and posters with the salutation "Glory to Ukraine!" on them (Amazon, 2022; Saint Javelin, 2022). The salutation has also become a meme supporting Ukraine on platforms such as TikTok , Twitter and Facebook, and President Zelenskyy signs off his speeches with "Slavia Ukraini!". Several memes have been created featuring the blue and yellow flag with the words "Slava Ukraini!" and "Heroyam slava!" in Ukrainian and English versions (see Figure 5.13) (Challinor, 2022; Federal News Service, 2022; Garcia, 2022).

The Ukrainian army's adoption of the phrase as a military salute at the Independence Day parade in 2018 replaced the Soviet-era military greeting "Wish you health, comrade!" (Bang-Andersen, 2018; Kaniewski, 2018). Some critics have linked the salute to radical nationalism from the 1930s. However, during Euromaidan, the phrase lost its radical nationalistic connotation and took on a new meaning symbolising the new Ukraine (Shveda & Park, 2016, p. 90). As part of Kremlin propaganda, there have been attempts to spread rumours that the Euromaidan was a "Nazi" operation and that all of Ukraine's citizens were Nazis, in line with the claim to de-Nazify Ukraine (see Chapter 2). However, Russia has failed to gain support to link "Glory to Ukraine!" and "Glory to Heroes!" to Nazi Germany's "Heil Hitler" salute (Ukraine World, 2021).

Figure 5.13 Examples of Symbolism in Memes

Visual State Symbols

Symbols are embedded in various forms of visual communication, such as memes, and these memes can become symbols of certain events. All of these communication forms contribute to resistance against aggressors. They have been instrumental in building resilience and a shared sense of national and cultural identity, which may have been less prevalent in some areas during Russia's annexation of Crimea in 2014. Visuals can help audiences recognise something they have heard about or understand critical events, creating a strong link between communication and emotions. Visual images in memes enable viewers to quickly identify and relate symbols to their political stance. By seeing a visual image, the audience can easily link a political reality, event, and audience (in-groups and out-groups) depending on the argument they want to make using a particular image (Schill, 2012, p. 129).

Russia

The Letter Z

The letter Z represents the totalitarian state, symbolising military force and a sense of belonging (see Figure 5.14). Graphically, the Z is more similar to the Nazi symbol, the swastika, than any of the old Soviet symbols such as the five-pointed star, the hammer and sickle, or the red flower (KnowYourMeme, 2022e; Sauer, 2022a). The use of Z draws parallels to the use of the swastika in Nazi Germany before and during World War II (Butler, 2022; Greesen, 2022; KnowYourMeme, 2022e; Sauer, 2022a). While Russia has used other letters on its military equipment, including O, X, A, and V, Z remains the most well-known symbol (KnowYourMeme, 2022e; Sauer, 2022a). Several memes have been developed in Russia to support the military forces, using the Z and other symbols. When the Russian military entered Ukraine in February 2022, military vehicles and troops had a Z painted on the sides of their equipment (KnowYourMeme, 2022e; Sauer, 2022b).

The actual meaning of the letter Z on Russian military equipment remains contested. One popular theory is that the letter is linked to the territorial locations of the different Russian troops. For instance, Z could represent the Russian word for "the Western District" (ZVO), which translates as "the West" (Dean, 2022; European Times, 2022; Sauer, 2022b; Slisco, 2022). V is linked to military equipment and units from the Eastern District (VVO), which is further away from the conflict zone and does not appear as often as Z (European Times, 2022). The Russian Ministry of Defence has not commented on the sudden use of these letters. However, on its Instagram channel, the ministry claimed that Z stands for "za pobedu", which translates as "to victory", and V represents the "power of truth" (Dean, 2022; Sauer, 2022b; Slisco, 2022).

The Orange-Black Saint George Ribbon

The Orange-Black Saint George ribbon also symbolises the war. This matches the overwhelming use of Ukraine's blue and yellow colours (see Figure 5.15). The Order of Saint George is one of the few "tsarist" military awards still

Figure 5.14 The Symbol Z

in use today. Its legacy was protected during the Soviet Union, and Russia adopted it to symbolise its victory over Nazi Germany (Forces, 2021). The symbol is widely used with the Z to create a national political identity for Russia, matching Ukraine's successful use of a visually recognisable symbol. Although Russians have long neglected this symbol, it has regained popularity during the war with Ukraine. The black and orange striped ribbons are cynically capitalised to create a link to the "Great Patriotic War" (WWII) for political purposes. The ribbon has symbolic meaning for Russian proxies in eastern Ukraine, who began using it in support of Russia in 2014 (Slisco, 2022; Tanas, 2022). Hence it is now being used in conjunction with the Z, and related accessories can be purchased to support the war effort (Kolstø, 2016).

Moldova, Estonia, Latvia, and Lithuania have all banned the orange-black Saint George ribbon and the letter Z in response to Russia's aggression in Ukraine (Adric, 2022; Kyiv Independent, 2022a; Slisco, 2022; Tanas, 2022). The Ukrainian Minister of Foreign Affairs, Kuleba, has used the online environment to communicate this. Kuleba tweeted an appeal asking states to ban these symbols as they are associated with Russian war crimes (see Figure 5.20) (Kuleba, 2022). The online environment enables users to pressure their leaders, resulting in greater engagement in the problems of other countries. In support of Ukraine, there has been widespread outrage and calls for action, even among those without a direct national or personal connection to the issue (Economist, 2022b). This type of communication draws attention to the problem in a way that is not possible offline. The audience is more extensive, and it would be difficult for governments to ignore the rationale behind the ban.

Figure 5.15 The Ribbon of Saint George

I call on all states to criminalize the use of the 'Z' symbol as a way to publicly support Russia's war of aggression against Ukraine. 'Z' means Russian war crimes, bombed out cities, thousands of murdered Ukrainians. Public support of this barbarism must be forbidden.

Figure 5.16 Kuleba's Online Appeal to States

Ukraine

The Trident

Regarding state symbols, Ukraine has used its golden Tryzub [Trident] on a blue shield (see Figure 5.17). Its symbolic value has increased as it represents resilience in a broader context, linking the fight for independence 100 years ago with those fighting against Russian aggression to protect the sovereign state. The Trident became the symbol of the Ukrainian national movement for independence after the collapse of the Russian Empire (Cuksa, 2011; Luzan, 2022). The Trident was chosen in 1918 by the Ukrainian People's Republic government as the state coat of arms. After gaining independence in 1992, the Verkhovna Rada officially approved the Trident as a national emblem of Ukraine (Trach, 2016). The use of the Trident increased during the war as a way to symbolise resistance and solidarity. The symbol has been seen on military uniforms, flags, and in public demonstrations. It has also been widely used in social media campaigns to raise awareness and show support for Ukraine's struggle against Russian aggression.

The National Flag

Similar to the Trident, the Ukrainian flag holds significant symbolic value. Its recognisability outside Ukraine is due to its constant exposure during the war and its association with solidarity. The Ukrainian flag has two horizontal parts, with blue on the top representing the sky and yellow on the bottom representing the grains growing in wheat fields, signifying the country's role as Europe's breadbasket (Flag Institute, 2022). Since the Russian invasion in 2022, the colours of the Ukrainian flag have been displayed everywhere,

Figure 5.17 The Ukrainian Coat of Arms: The Trident

from light projections on landmarks to buildings and clothing, as the blue and yellow have become symbols of solidarity and quiet resistance (Brandon, 2022; Lakritz, 2022; William, 2022). European Commission President von der Leyen is often dressed in these two colours, such as in her address to the European Parliament when she recommended that Ukraine join the EU (Epstein, 2022; Politico, 2022). Members of the US Congress also sent a message to Ukraine from an early stage. During President Biden's State of the Union address in March 2022, many Congress members wore yellow and blue, including the First Lady, Jill Biden (Lakritz, 2022; Pereira, 2022b). President Zelenskyy's first wartime visit to Washington (2022) was packed with symbolism. President Zelenskyy presented a flag signed by Ukrainian troops at the frontline in Bakhmut to the Speaker of the House and the Vice President during the Joint Meeting in the US Congress (Knox & Anders, 2022; Liptak, 2022). The flag and the Trident are used in various forms of communication and memes, serving as symbols that promote civic identity and collective memory of the situation in Ukraine. These symbols are crucial for cultural, societal, and political identities and for promoting inclusivity within the country. Meanwhile, outside Ukraine, they represent the fight against an aggressor, bravery, and hope (see Figures 5.18–5.20) (Oswald et al., 2022).

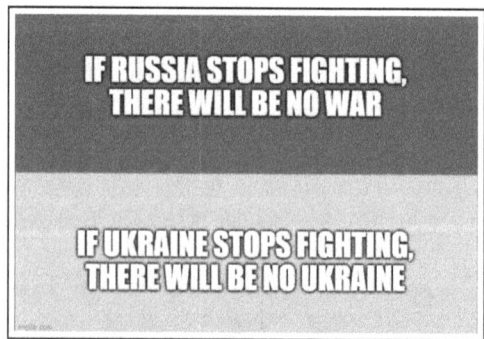

Figure 5.18 Meme with Ukrainian Symbol (a)

Figure 5.19 Meme with Ukrainian Symbol (b)

Figure 5.20 Meme with Ukrainian Symbol (c)

Other Symbols

Other symbols are associated with the war and Ukraine's civic stance against Russia. These symbols follow the same trend of centring Ukraine and encouraging people to unite in a shared understanding of people's hardships and difficulties. They also serve as defined symbols of resistance and civic engagement. These events have been incorporated into memes that are widely circulated beyond the country. The stories behind the memes and their symbolic status are now part of Ukrainian war folklore. These symbols reveal the context for what the meme tries to accomplish without further explanation. Showing a Ukrainian flag or the Trident gives the viewer an instant understanding that this communication is rooted in the Ukrainian war. Both verbal and non-verbal memes have distinctive characteristics that are essential for the viewer to analyse to get the intended outcome (Grundlingh, 2018, p. 154).

Sunflower

The sunflower's symbolic value has grown since the war started (Hassan, 2022; Waxman, 2022). Sunflowers have emerged as a national symbol of peace and resistance and are often used to show solidarity with Ukraine. However, they are also used to mock Russian soldiers and their likely fate of fighting on Ukrainian soil (see Figure 5.21). A viral video on Twitter showed a Ukrainian woman in Henychesk offering sunflower seeds to a Russian soldier after he ignored her question about their presence in Ukraine. As she handed him the sunflowers, she said:

> Take these seeds and put them in your pockets, so at least sunflowers will grow when you all lie down here. (Hassan, 2022; Sommerlad, 2022; Waxman, 2022)

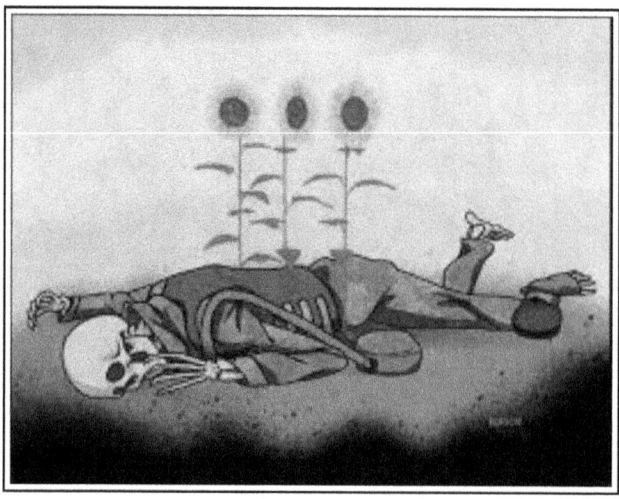

Figure 5.21 The Sunflower: Symbol of Dead Russian Soldiers

In response to the situation in Ukraine, artists worldwide have shared sunflower-themed art using hashtags such as #sunflowersforukraine (see Figure 5.22). On Twitter and Instagram, users have been adding sunflower emojis to their usernames alongside the Ukrainian flag as a sign of support. These memes and communications demonstrate solidarity with the Ukrainian people through powerful displays of artistic expression (Hassan, 2022; Joyner, 2022).

Symbols such as the sunflower and the flag have gained global momentum through their exposure on social media. For instance, at the State of the Union event, US First Lady, Jill Biden, had a sunflower embroidered on the sleeve of her dress, and she was also photographed wearing a sunflower-decorated face mask (Givhan, 2022; Hassan, 2022; Joyner, 2022; Waxman, 2022). These symbols are associated with the softer human side of the war. Memes and online visual and non-visual communications keep reminding people and help people internally and externally to decode messages and memes. Although these symbols do not hold the same significance in war as weapons do, they add another dimension to ensure that support for Ukraine does not diminish as the war drags on (Sabbagh, 2022; Tsakiris, 2022). Support for Ukraine is crucial as it is linked to the need for a constant flow of money, support, and military equipment to make military progress. If the feeling of solidarity disappears, economic support may also dry up or decrease.

Figure 5.22 Artwork Symbolising Ukraine

Watermelon

Other symbols have also emerged as equally powerful to create the feeling of solidarity, social identities and experiences (Keblusek et al., 2017, p. 1). The watermelon symbolises the victory of the Ukrainian armed forces and civilians over the occupation in the Kherson region. It has become a mascot for the city and region's struggle for freedom (see Figure 5.23) (Bubalo, 2022; We are Ukraine, 2022). The watermelon's association with the Kherson region has given it a special significance in the context of the war in Ukraine. Additionally, President Zelenskyy added to the symbolic status of the watermelon by jokingly claiming that he only visited Kherson after the liberation because he wanted to have some watermelon (Bubalo, 2022; Economist, 2022a).

Figure 5.23 Symbol on the Liberation of Kherson

In the digital age, symbols and codes have become even more critical for communication and creating shared meanings. They allow for the quick and efficient transmission of ideas and can evoke powerful emotions and associations. Symbols can also be a shorthand for complex ideas, allowing for more straightforward and concise communication. Symbols can have meaning beyond their literal definition, with social interactions and shared experiences shaping their significance. In the case of Ukraine, symbols such as the sunflower and the watermelon have become powerful tools for expressing solidarity and resistance – and defining Ukraine's cultural and political identity (Athique, 2015, p. 97).

Memes and Key Events

There are various reasons why memes are created, and the events of the war, including victories and losses, are often a part of it. Memes can be seen as a way of lashing out at Russia and the Russian army. The concept of a memetic war includes political expression and participation without physically being in the country or on the frontlines. Social media platforms facilitate the rapid spread of information, encouraging memesters and online users to express their opinions and become active participants in the process. Political humour is a significant aspect of this communication, focusing on areas such as political issues, people, events, processes, and institutions (Economist, 2022b; Young, 2018, pp. 872–873). Decoding memes requires a shared understanding of their intended purpose and the context in which they are used. Both verbal and non-verbal memes have distinct characteristics that viewers must analyse to understand their intended meaning (Grundlingh, 2018, p. 154).

Social identity is a central element in the world of memes, where different groups are represented, and emotions such as anger, sadness, and happiness are incorporated into the culture (Nissenbaum & Shifman, 2018). Unlike in conventional wars, social media have played an unprecedented role in disseminating information during the memetic war, where governmental and civil online users engage a large audience (Economist, 2022b). Inspirational and supportive memes are part of the memetic war, where verbal and non-verbal memes are used to support the in-group and help the Ukrainian population to overcome difficult times (see Figures 5.24–5.26) (see Chapter 4). The consistent messaging strategy from the Presidential office and other governmental and armed forces actors helps circulate a coherent message picked up by memesters and online users. The frequent use of symbols supports this approach by making communication more recognisable.

Figure 5.24 Example of a Positive and Supportive Meme

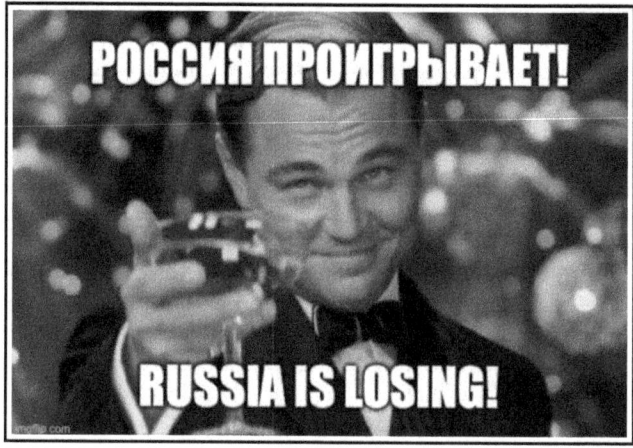

Figure 5.25 Meme with a Message to Russia

Figure 5.26 Meme with Link to the Strength of Ukraine

Anti-Russian Memes

Russian propaganda has been the subject of many memes on various topics. Memes related to the Russian-Ukrainian war spread across several platforms around mid-January 2022, including issues related to the brewing conflict before it became a full-scale war (KnowYourMeme, 2022). The info war quickly turned into a meme war, with different topics used to ridicule Russia. There are numerous themes for the memes. However, the following sections will include selected meme campaigns that went viral on a large scale. These should provide valuable insight into the types of meme developed to mock Russia and support Ukraine. Some meme campaigns have been outlined in previous sections, so they will not be repeated. However, a theme emerges, setting a clear distinction between the in-group and the out-group, creating a common purpose and a feeling of togetherness that amplifies the resistance against the Russian aggressors. Themes are linked to atrocities, civilian losses, the designation of Russia as a terrorist state, Nazism, war crimes and war criminals, the absurdity of the Russian claims for invading Ukraine, and the special military operation (see Figures 5.27–5.30).

Figure 5.27 Meme Linking Russia to War Crimes and Attacks on Civilians

Figure 5.28 Meme Linking Russia to Terrorism

Figure 5.29 Meme Linking Russians to the War

Figure 5.30 Meme Proposing the End of Russia

Mocking the Russian Army

When the invasion started, Ukraine's official account encouraged the population to use Twitter and the online environment to create meme resistance:

Tag @Russia and tell them what you think about them. (Ukraine/Україна, 2022)

Several memes mocked President Putin and depicted him as Hitler or linked Nazi Germany with the Russian invasion of a sovereign state. These memes were reminiscent of a meme posted on Ukraine's government's official Twitter account, which featured a cartoon image of an adult Hitler looking adoringly down while striking the face of a small child-like Russian President Putin (see Figure 5.31) (Ukraine/Україна, 2022). The following meme from the account stated:

This is not a "meme", but our and your reality right now. (Ukraine/Україна, 2022)

Using memes has proved to be a powerful tool in countering Kremlin propaganda, dis-and-misinformation, and gaining international support. However, there was some initial criticism of Ukraine's use of memes during a time of war. Some argued that they should focus on the dangers facing the country instead. Nevertheless, in hindsight, memes have activated a force online and substantially helped counter Russia's propaganda and dis-and-misinformation campaign. Other common themes in memes include linking Russia to war crimes, Nazism, and terrorism (see Figures 5.32–5.34). The memes undermine the unsupported Nazi claim and highlight the absurdity of Russia's actions (see Chapter 3). Additionally, the memes mock the progress of the Russian Armed Forces, which were once considered the second-best army in the world before the 2022 full-scale invasion of Ukraine (see Figures 5.35, 5.36).

Figure 5.31 Meme Posted on Ukraine's Official Governmental Webpage

Figure 5.32 Memes with Link Between Hitler and President Putin

Figure 5.33 Meme Linking Russian War Crimes in Different Countries

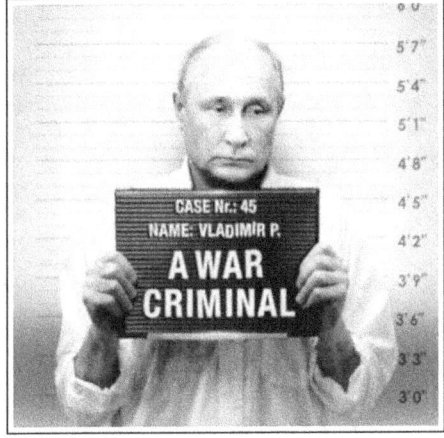

Figure 5.34 Meme Linking President Putin to the War Crimes Committed in Ukraine

Figure 5.35 Meme Mocking the Russian War Effort

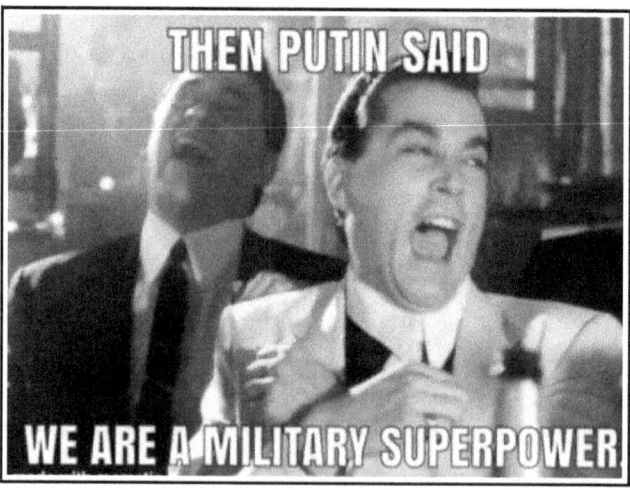

Figure 5.36 Meme Mocking the Russian Armed Forces

The Tractor Army

One of the most significant war memes to emerge is the tractor meme, which mocks the logistical problems faced by the Russian army. The meme began circulating early in the war after Russian tanks ran out of petrol and were abandoned by soldiers. Ukrainian farmers used their tractors to drag these tanks away, leading to online jokes about the "second strongest army in the world" losing equipment to farmers (Lutska, 2022). A short clip of a cartoon tractor dragging a Russian tank with a big Z on the side went viral on TikTok, with hashtags such as #IStandWithUkraine and #SlavaUkraini (Moody, 2022; Walfisz, 2022). The hashtag #Ukranian #tractors also gained popularity, fuelling an overwhelming creativity in developing memes with the trademark tractor (see Figure 5.37) (Hemsworth, 2022). These memes, such as the Ukrainian farmers vs Russian Army meme, have humiliated Russia. In addition, these memes have helped raise funds for charities with various causes, including supporting Ukraine's armed forces (Brzozowski, 2022; KnowYourMeme, 2022f).

Looting

The looting by Russian soldiers in Ukraine has also become a meme topic, with washing machines being a central theme for the jokes. The soldiers looted the places they were staying in, stealing computers and other electronic equipment, as well as toilets, washing machines, and fridges, on an unprecedented

Figure 5.37 Example of a Ukrainian Tractor Meme

scale (Walker & Roth, 2022). Numerous videos and images have been posted online showing Russian soldiers looting and sending the stolen material back to Russia (see Figure 5.42) (Golder & Murray, 2022; Jewers, 2022; Kika, 2022). Footage from Belarus shows that the soldiers have sent back hundreds of kilograms of stolen goods. The inspiration for the memes derives from the irony that Russian soldiers loot items such as toasters or scooters as part of their special operation to "save and liberate" (see Figure 5.38) (Lutska, 2022; UN News, 2022).

President Zelenskyy has addressed the issue of looting several times in his speeches to the Ukrainian people, which has led to the creation of new memes. This coordination between speeches and memes helps to keep the issue of looting and other behaviours by the Russian army fresh in people's minds. In a speech on 5 January 2023, President Zelenskyy said:

> So, let them take the toilet bowls – they'll need them on the road – and go back home. Behind our border of 1991. (President of Ukraine, 2023)

How to catch a
Russian soldier

Figure 5.38 Meme Linked to Russians' Looting in Ukraine

Mocking President Putin

Mocking President Putin is a recurring theme, with numerous memes poking fun at his isolation and his habit of keeping a physical distance from others. One frequently reused meme theme is Putin's long table. The meme refers to an image of a meeting between Putin and French President Macron before the 2022 full-scale Russian invasion, where the two leaders were placed at opposite ends of a very long table. The absurdity of this setup did not go unnoticed, and memes appeared to mock the long table and the distance between the leaders. The image was edited to portray the two men in different situations, such as playing badminton or having a magical laser beam battle (see Figure 5.39) (Bishara, 2022a; KnowYourMeme, 2022d). Another meme-worthy long table was used in separate meetings between Putin and his two ministers, Foreign Minister Lavrov and Defence Minister Shoigu, placing them more than four metres apart (see Figure 5.40) (Sabin, 2022).

Figure 5.39 Meme Mocking President Putin's Long Table

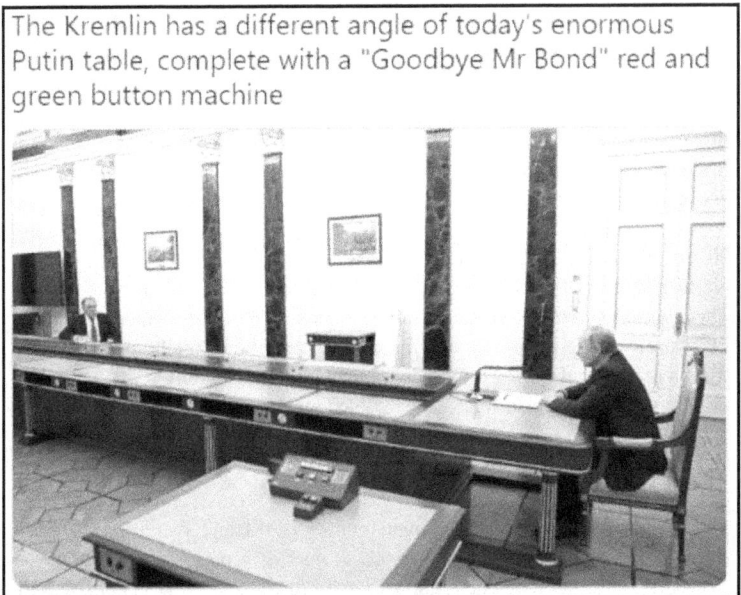

Figure 5.40 President Putin and Lavrov by the Long Table

Pro-Ukrainian Memes

As seen in political speeches, various techniques are used in memes to achieve different goals, such as ridiculing the enemy, boosting the morale of the civilian population or military forces, or gaining support for a particular argument. Emotions such as anger, sadness, and happiness are standard features of the meme culture (Nissenbaum & Shifman, 2018), as are other types of language, symbol, and behaviour that are repeated and changed over time. Social identity is central to memes, with a significant part of the culture linked to representations of diverse groups. However, viewers can easily recognise which groups a meme belongs to, whether it is the negative out-group or the positive in-group. Pro-Ukraine memes cover three main themes. First, they boost the country's morale and praise Ukrainian soldiers for their defence. Second, the memes mock the Russian troops for their lack of skills and President Putin for his decisions related to the war. Last, the memes criticise the West for its lack of help and fear of escalation, such as the United Nations and NATO (Bishara, 2022b).

Victories

Moskva Sinking

Another popular meme is the "Moskva sinking", which refers to the Russian warship from the Black Sea fleet that allegedly was attacked by Ukrainian forces in April 2022. Russia initially tried to deny that the ship was attacked, but it later sank due to a fire on board. This was the most significant wartime loss of a naval vessel in 40 years (see Figure 5.41) (KnowYourMeme, 2022c). In October, Ukraine had another maritime success. Online footage showed a drone boat attacking Russian ships at Sevastopol port in Crimea. The drone appeared to have hit frigate *Admiral Makarov* and a minesweeper (Guardian, 2022b; Niemeyer, 2022). Videos of resistance demonstrate that it is possible to take down tanks, sink battleships, and defend against military power. Through online communications and the constant flow of memes ridiculing the war effort, the world is aware that a large part of the communication from Russia is false and propaganda. These memes serve to highlight the successes of Ukraine and diminish Russia's perceived strength (Pereira, 2022a).

Snake Island

The exchange between a Ukrainian border guard and two Russian warships at Snake Island has become the subject of a popular meme. On 24 February 2022, the Russian warships *Vasily Bykov* and *Moskva* approached Snake Island and demanded that the Ukrainian border guards stationed there

Figure 5.41 Meme of *Moskva* Sinking

surrender or face being bombed. In response, a Ukrainian soldier stationed on the island famously replied with this pithy comment (Visontay, 2022; Walfisz, 2022):

> Russian warship, go fuck yourself. (Harding, 2022; Sheftalovich, 2022; Visontay, 2022)

This event quickly became a symbol of resistance and inspired various forms of communication, including postage stamps and memes (see Figure 5.42) (KnowYourMeme, 2022a; Parker, 2022; Pereira, 2022a). The meme served its purpose of boosting the morale of the Ukrainian population, and it became a symbol of resistance against the Russian invasion.

The Ghost of Kyiv

The Ghost of Kyiv meme has been used to create and promote a shared experience in the face of adversity. The meme has gained momentum as a legendary tale of the Ukrainian resistance portraying a Ukrainian Falcon fighter

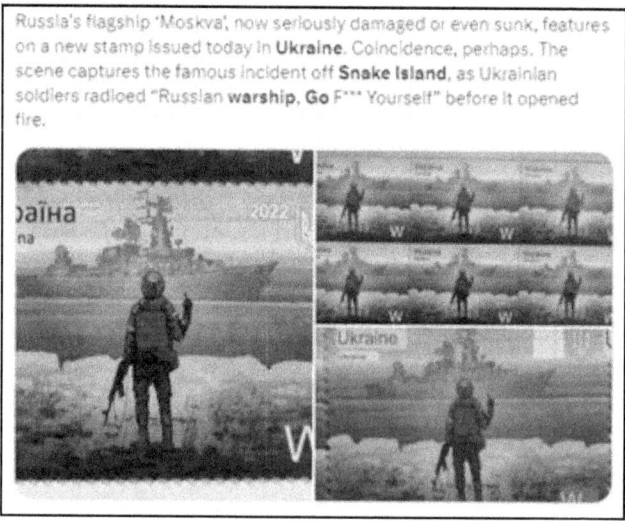

Figure 5.42 Postage Stamps Based on the Snake Island Incident

pilot flying a MiG-29, who allegedly took out six Russian fighter jets within 24 hours on 24 February 2022 (see Figure 5.42) (KnowYourMeme, 2022b). Conflicting reports emerged about the validity of this action, with different stories confirming and rejecting the Ghost of Kyiv meme (Beachum, 2022; Brown, 2022; Dress, 2022; KnowYourMeme, 2022b; Kyiv Post, 2022; New York Post, 2022a, 2022b;). The meme has also been used to create new content, such as artwork and merchandise, prominently featuring the Ghost of Kyiv image (Saint Javelin, 2022). Even if the story of the Ghost of Kyiv is not entirely true, the impact of the meme on the morale of the Ukrainian people and their supporters is undeniable.

Kerch Bridge

On October 2022, a section of the Kerch bridge collapsed after a blast. This attack was a significant blow to Russian military prestige and a big win for the Ukrainian side, although Ukraine never took credit for the attack (Beaumont, 2022; Beaumont & Graham-Harrison, 2022; Hayda, 2022). The attack on the Kerch bridge instantly generated several mocking memes to celebrate the success, with links to President Putin and his long table (see Figure 5.44) and his 70th birthday the day before the attack (see Figure 5.43). However, the reaction to the attack came promptly with a devastating number of missile strikes against Ukraine in response to what Putin deemed "Ukrainian terrorism" (Hayda, 2022).

Figure 5.43 Meme of the Ghost of Kyiv

Figure 5.44 Meme after the Kerch Bridge Attack (a)

Figure 5.45 Meme after the Kerch Bridge Attack (b)

References

Adric, L., 2022. *Estonia bans the display of the ribbon of Saint George and the letter "Z" as symbols of the Russian invasion.* [Online] Available at: https://www.royalsblue. com/estonia-bans-the-display-of-the-ribbon-of-saint-george-and-the-letter-z-as-symbols-of-the-russian-invasion/ [Accessed 07 01 2023].

Afanasiev, I., Mann, B., Selyukh, A., & Nadworny, E., 2022. *Ukraine agonises over Russian culture and language in its social fabric.* [Online] Available at: https://www. npr.org/ 2022/06/02/1101712731/russia-invasion-ukraine-russian-language-culture-identity [Accessed 29 12 2022].

Amazon, 2022. *Slava Ukraini.* [Online] Available at: https://www.amazon.com/slava-ukraini/s?k=slava+ukraini&rh=p_72%3A2661618011&dc&qid=1671391768&rnid =2661617011&ref=sr_nr_p_72_1&ds=v1%3AD2cTGnO%2B8lguGl0k4MJdzSrJz mKSG2LCBUsV4%2Blmzl4 [Accessed 18 12 2022].

Ascott, T., 2020. *How memes are becoming the new frontier of information warfare.* [Online] Available at: https://www.aspistrategist.org.au/how-memes-are-becoming-the-new-frontier-of-information-warfare/ [Accessed 25 12 2022].

Athique, A., 2015. *Digital Media and Society. An Introduction.* 2nd ed. Cambridge: Polity Press.

Bang-Andersen, C., 2018. *Glory to Ukraine? The de-glorified truth of Ukrainian nationalist policy.* [Online] Available at: https://georgetownsecuritystudiesreview. org/2018/10/22/glory-to-ukraine-the-de-glorified-truth-of-ukrainian-nationalist-policy/ [Accessed 18 12 2022].

Barnes, K., Riesenmy, T., Trinh, M.D., & Lleshi, E., 2021. Dank or Not? Analysing and Predicting. *Applied Network Science*, 6(1), pp. 1–24.

BBC News, 2022. *President Volodymyr Zelensky visits liberated Kherson.* [Online] Available at: https://www.bbc.co.uk/news/av/world-63626081 [Accessed 01 01 2023].

Beachum, L., 2022. *The "Ghost of Kyiv" was never alive, Ukrainian air force says.* [Online] Available at: https://www.washingtonpost.com/world/2022/05/01/ghost-of-kyiv-propaganda/ [Accessed 17 09 2022].

Beaumont, P., 2022. *Impact of Kerch bridge blast will be felt all the way to the Kremlin.* [Online] Available at: https://www.theguardian.com/world/2022/oct/08/impact-of-kerch-bridge-blast-will-be-felt-all-the-way-to-the-kremlin [Accessed 01 01 2023].

Beaumont, P. & Graham-Harrison, E., 2022. *Key bridge linking Crimea to Russia hit by huge explosion.* [Online] Available at: https://www.theguardian.com/world/2022/ oct/08/ crimea-kerch-bridge-explosions-russia-ukraine [Accessed 01 01 2023].

Bishara, H., 2022a. *The most hilarious memes about Putin and Macron's bizarre Kremlin meeting.* [Online] Available at: https://hyperallergic.com/710172/the-most-hilarious-memes-about-putin-and-macrons-bizarre-kremlin-meeting/ [Accessed 01 01 2023].

Bishara, H., 2022b. *Ukrainians wage a meme war against Russia.* [Online] Available at: https://hyperallergic.com/716738/ukrainians-wage-a-meme-war-against-russia/ [Accessed 01 01 2023].

Borger, J., 2022. *Much of Ukraine still without power, heat and water after missile attacks.* [Online] Available at: https://www.theguardian.com/world/2022/nov/25/much-of-ukraine-without-power-heat-and-water-after-missile-attacks [Accessed 30 11 2022].

Braithwaite, S., 2022. *Zelensky refuses US offer to evacuate, saying "I need ammunition, not a ride".* [Online] Available at: https://edition.cnn.com/2022/02/26/europe/ ukraine-zelensky-evacuation-intl/index.html [Accessed 25 12 2022].

Brandon, E. M., 2022. *In a visual rebuke to Putin, Ukraine's colors are being displayed in protest all over the world.* [Online] Available at: https://www.fastcompany.com/ 90732077/in-a-visual-rebuke-to-putin-ukraines-colors-are-being-displayed-in-protest-all-over-the-world [Accessed 13 11 2022].

Brown, L., 2022. *Ghost of Kyiv is alive in all pilots fighting for Ukraine, says air force.* [Online] Available at: https://www.thetimes.co.uk/article/ghost-of-kyiv-who-shot-down-more-than-40-russian-aircraft-dies-in-battle-q3sq0hztx [Accessed 17 09 2022].

Brzozowski, A., 2022. *Global Europe brief: How Ukraine is winning the meme war.* [Online] Available at: https://www.euractiv.com/section/europe-s-east/news/global-europe-brief-how-ukraine-is-winning-the-meme-war/ [Accessed 19 09 2022].

Bubalo, M., 2022. *Why did Zelensky want a watermelon in Kherson?.* [Online] Available at: https://www.bbc.co.uk/news/world-europe-63624456 [Accessed 23 11 2022].

Butler, M., 2022. *Ukraine's information war is winning hearts and minds in the West.* [Online] Available at: https://theconversation.com/ukraines-information-war-is-winning-hearts-and-minds-in-the-west-181892 [Accessed 19 09 2022].

Butz, D. A., 2009. National Symbols as Agents of Psychological and Social Change. *Political Psychology*, 30(05), pp. 779–804.

CBS News, 2022. *Putin calls opponents "scum and traitors" as Moscow announces new crackdown on "false information".* [Online] Available at: https://www.cbsnews.com/ news/putin-opponents-scum-traitors-repression/ [Accessed 21 03 2022].

Challinor, S., 2022. *Ukrainian president Zelensky receives standing ovation in Commons address.* [Online] Available at: https://www.leaderscouncil.co.uk/news/ukrainian-president-zelensky-receives-standing-ovation-in-commons-address [Accessed 18 12 2022].

Chraibi, C., 2016. *A short history of the Ukrainian greeting "Slava Ukrayini!".* [Online] Available at: https://euromaidanpress.com/2016/06/13/a-short-history-of-the-ukrainian-greeting-slava-ukrayini/ [Accessed 28 12 2022].

Dean, J., 2022. *The letter Z is becoming a symbol of Russia's war in Ukraine. But what does it mean?* [Online] Available at: https://www.npr.org/2022/03/09/1085471200/the-letter-z-russia-ukraine [Accessed 11 11 2022].

Demchak, C. C., 2011. *Wars of Disruption and Resilience.* Athens: University of Georgia Press.

Dress, B., 2022. *Video shows Ukrainian fighter jet shot down over Kyiv, officials say.* [Online] Available at: https://thehill.com/policy/international/595816-video-shows-ukrainian-fighter-jet-shot-down-over-kyiv-officials-say/ [Accessed 17 09 2022].

Economist, The, 2022a. *Kherson celebrates liberation and a visit from Volodymyr Zelensky.* [Online] Available at: https://www.economist.com/europe/2022/11/14/kherson-celebrates-liberation-and-a-visit-from-volodymyr-zelensky [Accessed 11 03 2023].

Economist, The, 2022b. *Ukraine's meme war with Russia is no laughing matter.* [Online] Available at: https://www.economist.com/europe/2022/03/01/ukraines-meme-war-with-russia-is-no-laughing-matter [Accessed 15 03 2023].

Elgenius, G., 2011. *Symbols of Nations and Nationalism.* 1st ed. Basingstoke: Palgrave Macmillan.

Ellyatt, H., 2022. *"There may be no light for a very long time": Ukrainians face a massive test of survival this winter.* [Online] Available at: https://www.cnbc.com/2022/11/22/ ukrainians-faces-a-dark-cold-winter-testing-their-resilience.html [Accessed 30 11 2022].

Epstein, J., 2022. *European Commission president wore an outfit in the colors of Ukraine's flag as she recommended the nation join the EU.* [Online] Available at: https://www. businessinsider.com/european-union-official-ukraine-colors-recommended-bloc-russia-war-2022-6?r=US&IR=T [Accessed 08 01 2023].

European Times, The, 2022. *What do the curious letters "Z" or "V" painted on Russian tanks mean?* [Online] Available at: https://www.europeantimes.news/2022/05/what-do-the-curious-letters-z-or-v-painted-on-russian-tanks-mean/#:~:text=   Although% 20his%20name%20in%20Cyrillic,   from%20the%20current%   20conflict%20zone [Accessed 11 11 2022].

Farkas, X. & Bene, M., 2021. Images, Politicians, and Social Media: Patterns and Effects of Politicians' Image-Based Political Communication Strategies on Social Media. *International Journal of Press/Politics*, 26(1), pp. 119–142.

Federal News Service, 2022. *Text of Ukrainian President Zelensky's virtual address to Congress.* [Online] Available at: https://www.washingtonpost.com/politics/ 2022/03/16/text-zelensky-address-congress/ [Accessed 18 12 2022].

Flag Institute, 2022. *Flag of Ukraine.* [Online] Available at: https://www.flaginstitute. org/wp/ 2022/03/flag-of-ukraine/#:~:text=The%20blue%20 and%20the%20yel-low%20of% 20Ukraine's%20flag%20represent%20the, in%20the%20huge% 20 wheatfields%20beneath [Accessed 13 11 2022].

Floyd, K., Schrodt, P., Erbert, L.A., & Scharp, K.M., 2022. *Exploring Communication Theory.* 2nd ed. New York: Routledge.

Forces, 2021. *The Ribbon of Saint George: Russia's version of rhe Remembrance poppy.* [Online] Available at: https://www.forces.net/russia/ribbon-saint-george-russias-version-remembrance-poppy [Accessed 07 01 2023].

Galey, R., 2022. *Ukraine admits the "Ghost of Kyiv" isn't real, but the myth was potent for a reason.* [Online] Available at: https://www.nbcnews.com/news/world/ukraine-admits-ghost-kyiv-isnt-real-wartime-myth-russia-rcna26867 [Accessed 20 09 2022].

Garcia, E., 2022. *Congress for no-fly zone in stirring speech tying US and Ukrain-ian patriotism.* [Online] Available at: https://www.independent.co.uk/news/world/ americas/us-politics/zelensky-speech-congress-today-ukraine-b2037281.html [Ac-cessed 18 12 2022].

Guardian, The, 2022a. *Russian state TV cuts away from Putin at pro-Russia rally – video.* [Online] Available at: https://www.youtube.com/watch?v=kp8qLWbxVTQ [Accessed 07 01 2022].

Guardian, The, 2022b. *Footage appears to show drone boat attack on Russian ships in Crimea – video.* [Online] Available at: https://www.theguardian.com/world/ video/ 2022/oct/30/footage-appears-to-show-drone-boat-attack-on-russian-ships-in-crimea-video [Accessed 07 01 2022].

Genauer, J., 2022. *A struggle between normality and madness: Why Volodymyr Zelensky's speeches have captured the world's attention.* [Online] Available at: https://theconver-sation.com/a-struggle-between-normality-and-madness-why-volodymyr-zelenskys-speeches-have-captured-the-worlds-attention-193224 [Accessed 29 12 2022].

Givhan, R., 2022. *The limits of performative symbolism.* [Online] Available at: https:// www.washingtonpost.com/nation/2022/03/08/limits-performative-symbolism/ [Accessed 13 11 2022].

Golder, J. & Murray, G., 2022. *Thieving Russian soldiers caught leaving house in in-vaded Ukraine with washing machine.* [Online] Available at: https://www.mirror. co.uk/ news/world-news/thieving-russian-soldiers-caught-leaving-28299484 [Ac-cessed 30 11 2022].

Graber, D., 1981. Political Language. In: *Handbook of Political Communication.* Beverly Hills: Sage, pp. 195–223.

Greesen, M., 2022. *"Z" is the symbol of the new Russian politics of aggression.* [Online] Available at: https://www.newyorker.com/news/our-columnists/z-is-the-symbol-of-the-new-russian-politics-of-aggression [Accessed 19 09 2022].

Grundlingh, L., 2018. Memes as Speech Acts. *Social Semotics*, 28(2), pp. 147–168.

Haq, S.N., Nechyporenko, K., & Chernova, A., 2022. *"Without gas or without you? Without you": Zelensky's words for Russia as Ukraine sweeps through northeast.* [Online] Available at: https://edition.cnn.com/2022/09/12/europe/zelensky-message-kharkiv-russia-ukraine-intl/index.html [Accessed 19 09 2022].

Harding, L., 2022. *"Russian warship, go fuck yourself": What happened next to the Ukrainians defending Snake Island?* [Online] Available at: https://www.theguardian.com/world/2022/nov/19/russian-warship-go-fuck-yourself-ukraine-snake-island [Accessed 30 11 2022].

Hassan, J., 2022. *The sunflower, Ukraine's national flower, is becoming a global symbol of solidarity.* [Online] Available at: https://www.washingtonpost.com/ world/2022/03/02/ukraine-sunflower-solidarity-russia-war/ [Accessed 08 01 2023].

Hayda, J., 2022. *After the Crimean bridge attack, there are plenty of theories but few real answers.* [Online] Available at: https://www.npr.org/2022/10/13/1128625322/crimea-bridge-attack-theories [Accessed 01 01 2023].

Hemsworth, C., 2022. *@_ChrisHemsworth.* [Online] Available at: https://twitter.com/_ChrisHemsworth/status/1508415341818482689 [Accessed 30 11 2022].

Holmes, O., 2022. *Putin's massive table: Powerplay or paranoia?.* [Online] Available at: https://www.theguardian.com/world/2022/feb/08/vladimir-putin-massive-table [Accessed 20 03 2022].

Human Rights Watch, 2022. *Ukraine: Russian forces' trail of death in Bucha.* [Online] Available at: https://www.hrw.org/news/2022/04/21/ukraine-russian-forces-trail-death-bucha [Accessed 03 03 2023].

Jewers, C., 2022. *Now Russia "liberates"... a WASHING MACHINE: Putin's men are spotted stealing kitchen appliance from Ukrainian home.* [Online] Available at: https://www.dailymail.co.uk/news/article-11340285/Putins-men-spotted-stealing-WASHING-MACHINE-Ukrainian-home.html [Accessed 30 11 2022].

Joyner, L., 2022. *Ukraine's sunflower becomes worldwide symbol of solidarity and peace amid Russian invasion.* [Online] Available at: https://www.housebeautiful.com/ uk/lifestyle/a39384516/russia-ukraine-war-sunflower-solidarity/ [Accessed 08 01 2023].

Kaniewski, D., 2018. *"Glory to Ukraine" army chant invokes nationalist past.* [Online] Available at: https://www.dw.com/en/new-glory-to-ukraine-army-chant-invokes-nationalist-past/a-45215538 [Accessed 18 12 2022].

Keblusek, L., Giles, H., & Maass, A., 2017. Communication and Group Life: How language and symbols shape intergroup relations. *Group Processes and Intergroup Relations*, 20(5), pp. 1–12.

Kestenholz, D., 2023. *Putin surrounds himself with extras instead of soldiers.* [Online] Available at: https://switzerlandtimes.ch/world/putin-surrounds-himself-with-extras-instead-of-soldiers/ [Accessed 01 01 2023].

Khurshudyan, I., 2022. *Ukraine's showdown with Russia plays out one meme at a time.* [Online] Available at: https://www.washingtonpost.com/world/2022/01/26/ukraine-russia-memes-social/ [Accessed 15 03 2023].

Kiehart, P., 2022. *Photos: Liberated Kherson celebrates as Ukrainians prepare for an uncertain future.* [Online] Available at: https://www.npr.org/sections/pictureshow/2022/11/17/1137260765/kherson-ukraine-liberated-celebration-photos [Accessed 11 03 2023].

Kika, T., 2022. *Adam Kinzinger shares video of Russian soldiers allegedly mailing loot home.* [Online] Available at: https://www.newsweek.com/adam-kinzinger-shares-video-russian-soldiers-allegedly-mailing-loot-home-1694485 [Accessed 30 11 2022].

Kika, T., 2022. *Speculation swirls as Putin accused of using same woman in multiple photos.* [Online] Available at: https://www.newsweek.com/speculation-swirls-putin-accused-using-same-woman-multiple-photos-1770576 [Accessed 01 01 2023].

Knibbs, K., 2022. *Volodymyr Zelensky is not a meme.* [Online] Available at: https://www.wired.com/story/ukraine-volodymyr-zelensky-meme/ [Accessed 12 11 2022].

KnowYourMeme, 2022a. *Battle of Snake Island.* [Online] Available at: https://knowyourmeme.com/memes/events/battle-of-snake-island [Accessed 19 09 2022].

KnowYourMeme, 2022b. *Ghost of Kyiv.* [Online] Available at: https://knowyourmeme.com/memes/ghost-of-kyiv [Accessed 17 09 2022].

KnowYourMeme, 2022c. *Moskva sinking.* [Online] Available at: https://knowyourmeme.com/memes/events/moskva-sinking [Accessed 19 09 2022].

KnowYourMeme, 2022d. *Putin's long table.* [Online] Available at: https://knowyourmeme.com/memes/putins-long-table [Accessed 01 01 2023].

KnowYourMeme, 2022e. *Russian Z military symbol.* [Online] Available at: https://knowyourmeme.com/memes/russian-z-military-symbol [Accessed 07 01 2023].

KnowYourMeme, 2022f. *Ukrainian farmers vs. Russian Army.* [Online] Available at: https://knowyourmeme.com/memes/ukrainian-farmers-vs-russian-army [Accessed 19 09 2022].

Knox, O. & Anders, C., 2022. *Mr. Zelensky goes to Washington in trip packed with symbolism.* [Online] Available at: https://www.washingtonpost.com/politics/2022/12/21/mr-zelensky-goes-washington-trip-packed-with-symbolism/ [Accessed 08 01 2023].

Kolstø, P., 2016. Symbol of the War — But Which One? The St George Ribbon in Russian Nation-Building. *Slavonic and East European Review*, 94(4), pp. 660–701.

Kuklychev, Y., 2022. *Fact check: Russia claims massacre in Bucha "staged" by Ukraine.* [Online] Available at: https://www.newsweek.com/fact-check-russia-claims-massacre-bucha-staged-ukraine-1694804 [Accessed 12 02 2023].

Kuleba, D., 2022. *@DmytroKuleba.* [Online] Available at: https://twitter.com/DmytroKuleba/status/1508686459880062977?ref_src=twsrc%5Etfw%7Ctwcamp%5Etweetembed%7Ctwterm%5E1508686459880062977%7Ctwgr%5E04fb958edd375e1c692fedd7b81e5e31c3abab93%7Ctwcon%5Es1_&ref_url=https%3A%2F%2Fwww.newsweek.com%2Flithuania-protest [Accessed 07 01 2023].

Kyiv Independent, The, 2022a. *@KyivIndependent.* [Online] Available at: https://twitter.com/ kyivindependent/status/1514602252282957829?lang=en [Accessed 07 01 2023].

Kyiv Independent, The, 2022b. *Zelensky visits liberated Kherson.* [Online] Available at: https://kyivindependent.com/news-feed/reuters-zelensky-visits-liberated-kherson [Accessed 01 01 2023].

Kyiv Post, 2022. *Twitter.* [Online] Available at: https://twitter.com/KyivPost/status/1501924542259814407 [Accessed 17 09 2022].

Lakritz, T., 2022. *Lawmakers wore blue and yellow to the State of the Union in support of Ukraine.* [Online] Available at: https://www.insider.com/ukraine-blue-and-yellow-state-of-the-union-outfits-2022-3 [Accessed 08 01 2023].

Lavorgna, A., 2020. *Cybercrimes. Critical Issues in a Global Context.* London: Macmillan.

Leicester, J., Arhirova, H., & Mednick, S., 2022. *Bombed, not beaten: Ukraine's capital flips to survival mode.* [Online] Available at: https://apnews.com/article/russia-ukraine-kyiv-europe-moscow-power-outages-806a7657c0ce11d0f9054960eb5f825f [Accessed 07 01 2023].

Lindgren, S., 2022. *Digital Media and Society.* 2nd ed. London: Sage.

Lipschultz, J.H., 2022. *Social Media and Political Communication.* London: Taylor & Francis.

Liptak, K., 2022. *5 takeaways from Volodymyr Zelensky's historic visit to Washington.* [Online] Available at: https://edition.cnn.com/2022/12/21/politics/takeaways-volodymyr-zelensky-visit-to-washington/index.html [Accessed 08 01 2023].

Lutska, V., 2022. *Ukrainian meme forces: What makes us laugh in the times of Russia's invasion.* [Online] Available at: https://war.ukraine.ua/articles/ukrainian-meme-forces-what-makes-us-laugh-in-the-times-of-russia-s-invasion/ [Accessed 23 11 2022].

McNair, B., 2011. *An Introduction to Political Communication.* 5th ed. London: Routledge.

Mendel, J., 2022. *Once Ukraine returns to its occupied villages, so will life.* [Online] Available at: https://www.politico.eu/article/ukraine-return-occupied-village-life-russia-war/ [Accessed 30 11 2022].

Moody, M., 2022. *@_MJMoody_.* [Online] Available at: https://twitter.com/_MJ-Moody_/ status/1507711435723493380?ref_src=twsrc%5Etfw%7Ctwcamp%5Etw eetembed%7Ctwterm%5E1508913394396585985%7Ctwgr%5Ee96629cce0aee55d 5dd7719e49e349aa9cb1c36c%7Ctwcon%5Es3_&ref_url=https%3A%2F%2Fwww. euronews.com%2Fculture%2F2022%2F03% [Accessed 30 11 2022].

Mortensen, M. & Neumayer, C., 2021. The Playful Politics of Memes. *Information, Communication and Society,* 24(16), pp. 2367–2377.

Muldoon, O.T., Trew, K., & Devine, P., 2020. Flagging Difference: Identification and emotional responses to. *Journal of Applied Social Psychology,* 50, pp. 265–275.

Mulvey, S., 2022. *Ukraine's Volodymyr Zelensky: The comedian president who is rising to the moment.* [Online] Available at: https://www.bbc.co.uk/news/world-europe-59667938 [Accessed 20 03 2022].

Munk, T. & Ahmad, J., 2022. "I Need Ammunition, Not a Ride": The Ukrainian cyber war. *Comunicação e Sociedade,* 42, pp. 221–241.

Nechepurenko, I., 2022. *Putin holds a highly choreographed meeting with mothers of Russian servicemen.* [Online] Available at: https://www.nytimes.com/ 2022/11/25/ world/europe/putin-russia-soldiers-mothers-war.html [Accessed 01 01 2023].

New York Post, 2022a. *Twitter.* [Online] Available at: https://twitter.com/ nypost/status/1520068212867702790 [Accessed 17 09 2022].

New York Post, 2022b. *Twitter.* [Online] Available at: https://twitter.com/ nypost/status/1520831090377859072 [Accessed 17 09 2022].

Niemeyer, K., 2022. *Shocking video shows "massive" attack by drone boats targeting Russia's Black Sea fleet.* [Online] Available at: https://www.insider.com/shocking-video-shows-drone-boats-attack-on-russian-ships-2022-10 [Accessed 08 01 2023].

Onuch, O. & Hale, H.E., 2022. *The Zelensky effect.* London: Hurst.

Oswald, E.C., Esborg, L., & Pierroux, P., 2022. Memes, Youth and Memory Institutions. *Information, Communication and Society*, pp. 1–18.

Parker, C., 2022. *On Snake Island, defiant Ukrainians force a Russian withdrawal.* [Online] Available at: https://www.washingtonpost.com/world/2022/06/30/ukraine-snake-island-russian-withdrawal/ [Accessed 19 09 2022].

Pereira, I., 2022a. *Memes become weapons in Ukraine-Russia conflict.* [Online] Available at: https://abcnews.go.com/International/memes-weapons-ukraine-russia-conflict/story?id=83184578 [Accessed 19 09 2022].

Pereira, I., 2022b. *State of the Union awash in blue and yellow to support Ukraine.* [Online] Available at: https://abcnews.go.com/Politics/ukrainian-support-full-display-state-union/story?id=83184581 [Accessed 08 01 2023].

Politico, 2022. *As it happened: Ursula von der Leyen's State of the Union speech.* [Online] Available at: https://www.politico.eu/article/live-blog-ursula-von-der-leyens-state-of-the-union-speech/ [Accessed 08 01 2023].

President of Ukraine, 2022. *We stand, we fight and we will win. Because we are united. Ukraine, America and the entire free world – address by Volodymyr Zelenskyy in a joint meeting of the US Congress.* [Online] Available at: https://www.president.gov.ua/en/news/mi-stoyimo-boremos-i-vigrayemo-bo-mi-razom-ukrayina-amerika-80017 [Accessed 29 12 2022].

President of Ukraine, 2023. *The war will be over when Russian soldiers either leave or we drive them out – address by the President of Ukraine.* [Online] Available at: https://www.president.gov.ua/en/news/vijna-zakinchitsya-koli-rosijski-soldati-abo-pidut-abo-mi-yi-80249 [Accessed 08 01 2023].

Prothero, M., 2022. *"I need ammo, not a ride": How Ukraine's president rallied a nation.* [Online] Available at: https://www.vice.com/en/article/dyp54v/volodymyr-zelenskyy-ukraine-russia-kyiv [Accessed 25 12 2022].

Reuters, 2023. *Putin praises Russian Orthodox Church for backing troops in Ukraine.* [Online] Available at: https://www.usnews.com/news/world/articles/2023-01-06/putin-attends-orthodox-christmas-service-by-himself-in-kremlin [Accessed 07 01 2023].

Roth, A. & Sauer, P., 2022. *Putin talks to mothers of soldiers fighting in Ukraine in staged meeting.* [Online] Available at: https://www.theguardian.com/world/2022/nov/25/putin-talks-to-mothers-of-soldiers-fighting-in-ukraine-in-staged-meeting [Accessed 01 01 2023].

Sabbagh, D., 2022. *Ukraine fears western support will fade as media loses interest in the war.* [Online] Available at: https://www.theguardian.com/world/2022/jun/12/ukraine-fears-western-support-will-fade-as-media-loses-interest-in-the-war [Accessed 08 01 2023].

Sabin, L., 2022. *Vladimir Putin holding another meeting at extremely long table sparks jokes online.* [Online] Available at: https://www.independent.co.uk/news/world/europe/vladimir-putin-long-table-russia-ukraine-b2015099.html [Accessed 14 02 2023].

Saint Javelin, 2022. *SLAVA UKRAINI!.* [Online] Available at: https://www.saintjavelin.com/ collections/slava-ukraini [Accessed 18 12 2022].

Sauer, P., 2022a. *Putin praises Russian "unity" at rally as glitch cuts state TV broadcast.* [Online] Available at: https://www.theguardian.com/world/2022/mar/18/putin-praises-russian-unity-at-rally-but-state-tv-broadcast-is-cut-off [Accessed 20 03 2022].

Sauer, P., 2022b. *Why has the letter Z become the symbol of war for Russia?* [Online] Available at: https://www.theguardian.com/world/2022/mar/07/why-has-the-letter-z-become-the-symbol-of-war-for-russia [Accessed 07 03 2022].

Saul, D., 2022. *Putin's long tables rxplained: Why he puts some leaders, including Germany's Scholz, at an extreme distance.* [Online] Available at: https://www.forbes. com/sites/dereksaul/2022/02/15/putins-long-tables-explained-why-he-puts-some-leaders-including-germanys-scholz-at-an-extreme-distance/?sh=7fbc9d0d70fb [Accessed 20 03 2022].

Schill, D., 2012. The Visual Image and the Political Image: A review of visual communication research in the field of political communication. *Review of Communication*, 12(2), pp. 118–142.

Segal, E., 2022. *Why Zelensky's speech to Congress was a masterclass in crisis communication.* [Online] Available at: https://www.forbes.com/sites/ edwardsegal/ 2022/12/22/10-crisis-communication-strategies-and-tactics-zelensky-used-in-speech-to-congress/ [Accessed 29 12 2022].

Seib, P., 2021. *Information at War.* 1st ed. Cambridge: Polity Press.

Sheftalovich, Z., 2022. *"Go fuck yourself", Ukrainian soldiers on Snake Island tell Russian ship before being killed.* [Online] Available at: https://www.politico.eu/ article/go-fuck-yourself-ukraine-soldiers-snake-island-russia-war-ship-killed/ [Accessed 30 11 2022].

Shveda, Y. & Park, J., 2016. Ukraine's Revolution of dignity: The dynamics of Euromaidan. *Journal of Euroasian Studies*, 7, pp. 85–91.

Sky News, 2022. *@SkyNews.* [Online] Available at: https://twitter.com/skynews/status/ 1498631433392377856?lang=en-GB [Accessed 25 12 2022].

Slisco, A., 2022. *Lithuania Protests Russian Aggression by Banning Display of letter "Z".* [Online] Available at: https://www.newsweek.com/lithuania-protests-russian-aggression-banning-display-letter-z-1699151 [Accessed 07 01 2023].

Smith, A., 2022. *"Scum and traitors": Under pressure over Ukraine, Putin turns his ire on Russians.* [Online] Available at: https://www.nbcnews.com/news/world/ scum-traitors-pressure-ukraine-putin-turns-ire-russians-rcna20410 [Accessed 21 03 2022].

Sommerlad, J., 2022. *What is the national flower of Ukraine?* [Online] Available at: https:// www.independent.co.uk/news/world/europe/ukraine-what-is-national-flower-sunflower-b2054864.html [Accessed 23 11 2022].

Stupples, D., 2015. *What is information warfare?* [Online] Available at: https://www. weforum.org/agenda/2015/12/what-is-information-warfare/ [Accessed 26 12 2022].

Sujon, Z., 2021. *The Social Media Age.* 1st ed. London: Sage.

Sussex, M., 2022. *With his army on the back foot, is escalation over Ukraine Vladimir Putin's only real option?.* [Online] Available at: https://theconversation.com/with-his-army-on-the-back-foot-is-escalation-over-ukraine-vladimir-putins-only-real-option-190046 [Accessed 19 09 2022].

Tanas, A., 2022. *Moldova parliament bans pro-Russian ribbon despite opposition walk-out.* [Online] Available at: https://www.reuters.com/world/europe/moldova-parliament-bans-pro-russian-ribbon-despite-opposition-walk-out-2022-04-07/ [Accessed 07 01 2023].

TikTok, 2022. *Ukraine slava ukraini meme.* [Online] Available at: https://www.tiktok. com/discover/Ukraine-slava-ukraini-meme [Accessed 18 12 2022].

Trach, N., 2016. *The story behind 2 top Ukrainian symbols: National flag and Trident.* [Online] Available at: https://www.kyivpost.com/article/content/ukraine-politics/ the-story-behind-2-top-ukrainian-symbols-national-flag-and-trident-421675.html [Accessed 08 01 2023].

Tsakiris, M., 2022. *Ukraine: How social media images from the ground could be affecting our response to the war.* [Online] Available at: https://theconversation.com/ukraine-how-social-media-images-from-the-ground-could-be-affecting-our-response-to-the-war-178722 [Accessed 08 01 2023].

Ukraine/Україна, 2022. *@Ukraine.* [Online] Available at: https://twitter.com/Ukraine/status/ 1496716168920547331?ref_src=twsrc%5Etfw%7Ctwcamp%5Etweetem bed%7Ctwterm%5E1496716168920547331%7Ctwgr%5E3ceb25ff92b43f20c57 920572d4fecfb636aeb4d%7Ctwcon%5Es1_&ref_url=https%3A%2F%2Fmetro.co.uk%2F2022%2F02%2F24%2Fukraine-se [Accessed 08 01 2023].

Ukraine World, 2021. *Why Kremlin propaganda lies about "Glory to Ukraine – Glory to Heroes" slogan.* [Online] Available at: https://ukraineworld.org/articles/ opinions/glory-ukraine-glory-heroes [Accessed 18 12 2022].

Ukrainian Memes Forces, 2022. *@uamemesforces.* [Online] Available at: https://mobile.twitter.com/uamemesforces/status/1609219802970587136 [Accessed 01 01 2023].

UN News, 2022. *Russia had "no choice" but to launch "special military operation" in Ukraine, Lavrov tells UN.* [Online] Available at: https://news.un.org/en/story/ 2022/09/1127881 [Accessed 30 11 2022].

Visontay, E., 2022. *Ukraine soldiers told Russian officer "go fuck yourself" before they died on island.* [Online] Available at: https://www.theguardian.com/world/2022/ feb/25/ukraine-soldiers-told-russians-to-go-fuck-yourself-before-black-sea-island-death [Accessed 30 11 2022].

Von Tunzelmann, A., 2022. *The big idea: Can social media change the course of war?.* [Online] Available at: https://www.theguardian.com/books/2022/apr/25/the-big-idea-can-social-media-change-the-course-of-war [Accessed 20 12 2022].

Walfisz, J., 2022. *Viral cartoon shows Ukrainian tractor dragging Russian tank: What's the story behind it?* [Online] Available at: https://www.euronews.com/culture/ 2022/03/30/viral-cartoon-shows-ukrainian-tractor-dragging-russian-tank-what-s-the-story-behind-it [Accessed 30 11 2022].

Walker, S., 2022. *Putin's absurd, angry spectacle will be a turning point in his long reign.* [Online] Available at: https://www.theguardian.com/world/2022/feb/21/putin-angry-spectacle-amounts-to-declaration-war-ukraine [Accessed 21 03 2022].

Walker, S. & Roth, A., 2022. *"They took our clothes": Ukrainians returning to looted homes.* [Online] Available at: https://www.theguardian.com/world/2022/apr/ 11/ ukrainian-homes-looted-by-russian-soldiers [Accessed 30 11 2022].

Waxman, O.B., 2022. *What to know about the meaning of sunflowers in Ukraine.* [Online] Available at: https://time.com/6154400/sunflowers-ukraine-history/ [Accessed 23 11 2022].

We are Ukraine, 2022. *How did the watermelon become a symbol of Kherson's liberation from Russian occupation?* [Online] Available at: https://www.weareukraine.info/ 38081-2/ [Accessed 23 11 2022].

William, H., 2022. *Landmarks turn yellow and blue in solidarity with Ukraine.* [Online] Available at: https://www.independent.co.uk/news/uk/ukraine-downing-street-boris-johnson-london-sergei-lavrov-b2023720.html [Accessed 08 01 2023].

Wood, M., 2022. *Everything Is Possible.* New York: Holland Publishing.

Yar, M. & Steinmetz, K.F., 2019. *Cybercrime and Society.* 3rd ed. London: Sage.

Young, D.G., 2018. Theories and Effects of Political Humor: Discounting cues, gateways, and the impact of incongruities. In: *The Oxford Handbook of Political Communication.* Oxford: Oxford University Press, pp. 871–884.

Zelensky, V., 2022. *A Message from Ukraine.* London: Penguin Random House.

6 Concluding Remarks – Strategic Considerations

Tine Munk

Concluding Remarks

Online communication shapes public opinion and political relations in the 21st century, particularly with the emergence of new media such as the Internet and social media platforms. Online platforms are instrumental for communication in everyday life, creating bonds between users. In spite of the measures taken by Ukraine, the challenge of countering disinformation and propaganda remains significant, particularly in the age of social media and the spread of false information online. Ukraine had faced challenges in countering these threats for years, especially since 2014, when Russia annexed Crimea and invaded eastern Ukraine. Ukraine has developed a legal framework to counter Russia's disinformation and propaganda campaign (see Chapters 2 and 3). However, steps have been taken by creating institutions, promoting a legal framework, and raising awareness and education to improve digital literacy. Ukraine is also collaborating with external partners, such as the EU, to combat the propaganda and dis-and-misinformation (see Chapter 3).

The findings in the book show that an informal structure of memetic warfare has emerged online since the full-scale Russian war against Ukraine started on 24 February. The book's first part conceptualises the area by focusing on the war and critical events leading to the Russian invasion in February 2022. It is essential to understand the background of this illegal war, how Ukraine has developed as a sovereign state, and its challenges (see Chapters 1 and 2). As a part of the war, memetic warfare was introduced and positioned as cyberwar and information warfare. The use of hybrid war makes using the online environment and social media beneficial to reach out to a large group of people and distribute certain narratives. The focus on memetic warfare has previously been on the offensive use of memes (Hancock, 2010). However, this research links the concept to defensive use.

The book has used Ukraine as a case study due to the country's development of communication strategies, the use of the online environment, and the networked actors and agencies. Several actors have progressed an informal meme strategy that aligns with the concept of memetic war. Using memes as

DOI: 10.4324/9781003432630-6

a defensive weapon is a new perspective currently being studied but not yet fully developed as an independent concept. This book introduces a framework for understanding the critical parameters to highlight key strategic points in the conclusion. The book's second part develops the concept of memetic war and tests it concerning actors and the use of memes. This part focuses on the different types of meme and what can be understood from studying them in the context of critical events. Chapter 3 establishes some essential parts of memetic warfare, further developed in two subsequent chapters. As a part of memetic warfare, Chapter 4 investigates the actors' roles, and Chapter 5 identifies the different types of communication, sign, and language used to produce memes. The areas identified in the memes also have a solid link to governmental communications and events during the war (see Chapter 5). Moreover, key themes assuch behaviours, language, and symbols are essential for creating a community of like-minded people. These memes are instrumental in developing a feeling of togetherness, strengthening the community, and fueling civic resistance (see Chapters 3, 4, and 5).

Key Findings from the Book

The book identifies the need for a comprehensive approach to memetic warfare, where defensive strategies are as important as offensive strategies. It emphasises the role of civic actors and grassroots organisations in countering false narratives and propaganda and how they can collaborate with government actors to achieve common goals. The book also highlights the importance of monitoring online activity and identifying patterns to develop effective strategies (see Chapters 3, 4, and 5).

Overall, the book provides valuable insights into the evolving landscape of memetic warfare and the role of the online environment in shaping public opinion and political relations. It underscores the need for a nuanced and multifaceted approach to memetic war that involves various stakeholders, including government actors, military actors, civic groups, and individuals. By understanding the underlying principles and dynamics of memetic warfare, countries can develop effective strategies to manage communications on social media and promote their national interests online.

The Concept of Memetic Warfare: A Strategic Approach

The chapters in this book develop the concept of memetic warfare holistically by outlining the context from which memes should be understood. The main focus on Ukraine highlights the country's success in leveraging the online environment to engage citizens and progress its narrative about the war. This success could serve as a model for other states looking to build resilience and resistance against potential threats. The book emphasises the importance of

Table 6.1 Descriptions of Key Areas

Key areas	Description of key areas
Complexity	• Memetic warfare is a multifaceted and intricate phenomenon encompassing various actors and motivations (Chapter 1) • Understanding the motivations and strategies of the diverse actors involved (Chapter 3) • Utilises different forms of communication to sway public opinion and bolster military forces (Chapter 5) • A non-linear and organic approach due to technological developments, digital literacies, a diverse group of actors, different types of behaviour and engagement • Ensure a prominent level of knowledge about the latest developments to adapt strategies accordingly (Chapter 1)
Community	• Developing a sense of community and fostering a supportive environment where individuals can protect and assist one another is crucial in the memetic warfare strategy (Chapter 4) • Involve a range of actors, including military combatants, civilians, private entities, and online activists (Chapters 3 and 4) • Skilled and non-skilled actors work alongside national and international actors and agencies in memetic warfare (Chapters 3 and 4)
Social Media	• Memes, videos, and other forms of online content are used to spread information and influence public opinion; a crucial aspect of this strategy enabling actors to reach a broad audience quickly, cheaply, and effectively (Chapters 1, 3, and 5) • Social media platforms have become increasingly prevalent; critical battlegrounds for competing narratives and ideologies (Chapters 2, 3, and 5) • Online communities are mobilising to generate support for various causes and resist online interference, propaganda, and dis-and-misinformation (Chapters 1, 3, and 5)
Public and Private Actors	• Memetic warfare requires collaboration between governmental, military, non-military private, and civilian groups and individuals who work together in both formal and informal settings (Chapter 3) • A successful memetic warfare strategy requires civic engagement and strategic governmental communications as key elements (Chapters 2 and 3) • The range of activities encompassed by memetic warfare is extensive, and the line between combatants and non-combatants is often indistinct (Chapters 2, 3, and 4) • Challenging to examine the specific roles of these two groups in greater detail. This highlights the need for further research to understand their respective functions better (Chapters 3 and 4)

community building and coordination among various actors to counter propaganda and dis-and-misinformation from hostile entities such as Russia (see Chapters 2 and 3) (Paul & Matthews, 2016). The spectrum of actions involved in memetic war includes a range of actors spanning from conventional warfare to civil activism. It involves a diverse mix of governmental, military, private non-military, and civilian groups and individuals working together (see Chapters 1, 2, and 3). From that point, the book provides valuable insights into the strategic challenges and potential benefits of leveraging the online environment for defensive memetic warfare.

The Legal Framework and Strategies

As memetic warfare involves a range of actors with various motivations, there is a need to establish clear roles and guidelines to prevent abuse and ensure ethical practices. Using private entities in memetic warfare may raise concerns about accountability and transparency, primarily when they operate outside traditional government structures. Additionally, as online communities play a crucial role in memetic warfare, protecting individuals' rights to free speech and privacy is necessary while countering propaganda and dis-and-misinformation.

The legal framework surrounding memetic warfare is still in its infancy, and international laws and norms may not adequately address the use of memes and other online communication forms in conflict situations (Munk, 2018). Therefore, it is necessary to develop new legal and ethical frameworks to guide memetic warfare actors and ensure that means and methods conform to established norms and values. Finally, it is essential to recognise that memetic warfare is but one tool in a broader toolkit for countering propaganda and dis-and-misinformation and engaging with civic society to build resilience against adversaries (Fiala, 2020; Fiala & Pettersson, 2020). Therefore, any memetic warfare strategy must be part of a broader, coordinated effort involving military, diplomatic, and civil society actors to achieve desired objectives.

There is currently no international framework to address the areas raised here. Further, treaties regulating the offline environment are unsuitable for addressing these issues (Munk, 2022, p. 115). While using memes in warfare may seem like a new phenomenon, it is addressed indirectly in the Tallinn Manual versions 1.0 and 2.0 (Schmitt, 2013, 2017). The Tallinn Manual 1.0 and 2.0 provide a guide on how international law applies to cyber operations covering areas such as the use of force, state responsibility, and the law of armed conflict (Munk, 2018, p. 239). In the context of memes, using this communication form in cyber operations might constitute a form of cyberattack, mainly if the memes are used to cause harm to a target's reputation or to spread false information. According to the manuals, memes might be considered a form of psychological operation, a tactic to influence an adversary's behaviour or beliefs. However, the defensive use of memes does not aim to harm or danger anyone (Schmitt, 2013, 2017). Instead, defensive memes minimise the damage of the adversary's communications by calling out falsehoods and providing information, so they have a different role. However, if using the offline rules for war and warfare, it is legal to engage in aggressive forms of defence to defend the nation from invading forces. However, the state cannot legally or ethically participate in preventive or preventive attacks (Manjikian, 2021, p. 289).

Key enablers of memetic warfare are closely tied to the national cybersecurity strategy and the different defensive cyber programmes developed on the governmental and military levels (Fiala, 2020, p. 115). The international society has failed to develop cyber-security treaties covering the use of the online environment in war and warfare, where two blocks of permanent states have blocked one another's initiatives (Munk, 2018, pp. 240–241; Munk, 2022, pp. 88–90). Most countries will have developed and implemented a national cybersecurity framework that covers areas such as protecting vital interests of the state and countering threats, harms, and damage from online attacks. Ukraine also has a well-developed framework; some initiatives are implemented, and others are being developed (CCDCOE, 2018). The essential parts are the Cyber Security Strategy of Ukraine 2016 (CCDCOE, 2018), and the Doctrine of Information Security of Ukraine 2017 (CoE, 2017). The military and cybersecurity strategies operate together, and the Strategic Defence Bulletin for Ukraine (President of Ukraine, 2021) is vital to consider. However, a grey area remains concerning the memetic warfare legal framework and its position within security strategies (see Chapters 2 and 3).

This implementation of security strategies and other initiatives to protect and defend cyberspace should be done to enhance resilience as a part of the country's Total Defence strategic foundation. Ukraine has been subjected to Russian offensive information warfare for a decade, where memetic communications are part of the strategy. The intensive use of memes to promote propaganda and dis-and-misinformation helps to confuse decision-makers and the general public, generate fear and cause uncertainty, confusion, and dissatisfaction with the government (Fiala, 2020, p. 115). The defensive memetic strategy has proven to be useful in countering Russia's offensive communications.

The Concept of ROC and Memetic Warfare

The resistance operational concept (ROC) is a military concept that aims to counter the conventional military superiority of an opposing force through irregular warfare tactics, such as guerrilla warfare, sabotage, and subversion. It involves using non-traditional tactics and strategies to resist and undermine the opponent's military and political objectives. By way of contrast, memetic warfare refers to using memes and other forms of cultural communication to spread ideas, influence, and shape attitudes and behaviour. It is often used in conjunction with other forms of information warfare, defined by national and military strategies.

The book examines the use of memes in support of the armed forces and governmental actors' cyberwar and information warfare initiatives in Ukraine. The concept of civic resistance is explored to understand how this memetic approach has evolved. The book also discusses the principles of the strategic military concept ROC, part of Ukraine's Total Defence Strategy, and the "whole-of-government" and "whole-of-society" approach that it advocates (see Chapter 2) (Fiala, 2020, p. 1). This strategy prepares governmental actors, military groups, civil groups, and individuals in peacetime to enable a comprehensive resistance movement during crises and war. In Ukraine, this strategy is implemented along with armed forces reforms, and the informal part of the ROC strategy can be observed in how governmental actors, civic groups, and individuals have developed a layered approach to online communications (see Chapters 2 and 3).

The book also highlights the emergence of groups forming a voluntary nodal structure outside the governmental framework, which share values with the government but act independently and decentralised. In the context of the ROC, memetic warfare can be a powerful tool for resistance forces to challenge the legitimacy and effectiveness of an opposing force (see Chapter 3). Using memes and other forms of cultural communication to disseminate information and ideas, resistance forces can create a narrative that undermines the opponent's objectives and weakens their support base. However, memetic warfare also has its limitations and potential drawbacks. Controlling the spread and interpretation of memes can be challenging, and they can be easily co-opted or used against the resistance forces (Mazarr et al., 2019; Iloh, 2021; Denisova, 2020, p. 32). Additionally, using memes and cultural communication can be seen as unconventional or unprofessional by some traditional military forces. While memetic warfare can be a valuable component of the ROC, it should be used strategically and in combination with other tactics and strategies to achieve the desired objectives (military and non-military cyber-strategies) (see Chapters 2 and 3).

Memetic Actions – Communication Strategy

It is important to note that the memetic warfare model used in Ukraine is based on a defensive approach to counter Russia's attempt to master the online environments, control the war narrative, and undermine the support for Ukraine. It can also effectively express frustration or anger towards a particular entity or group without violence or aggression. At the same time, Russia is pursuing a more offensive strategy, online and offline, attacking Ukraine, NATO, the EU, and the whole Western world using a combination of falsehoods, absurd arguments and examples, and information that is partly wrong or true information taken out of context.

The significance of memetic warfare and how it has become an essential tool for modern warfare are essential parts of this book (Chapters 3, 4, and 5). Memetic action involves the use of various communication forms. It means influencing public opinion and supporting military forces, including using social media and memes to counter propaganda and dis-and-misinformation.

The use of memes and memetic groups has proved successful as a part of the Ukrainian civic defence against Russian and pro-Russian actors online. The impact of the war in Ukraine has shown how social media can be used to raise awareness and garner support for a cause. At the same time, humour and satire can provide a sense of unity and resilience among people in times of conflict (Chapters 4 and 5).

Communications

It is essential to highlight the effectiveness of the communication strategy adopted by President Zelenskyy and key governmental actors and departments in Ukraine (Chapters 4 and 5). The strategy has successfully reached a large audience and gained momentum in the online war. The nodal structure formed around these communications contrasts with the Russian counterpart's centralised and state-centric structure. The analysis suggests replicating the informal memetic war structure can enhance resilience and ensure conflict resistance. However, the communication strategy must be clear, relevant, and consistent to uphold a decentralised and largely informal network. Without these attributes, online memetic forces lose momentum, interest, and engagement (see Chapters 3 and 4).

Furthermore, the defensive strategy's actions are beating Russia on its own turf by mirroring the Russian strategy defensively rather than offensively. This approach makes it increasingly tricky for pro-Russian actors online to manage online communications. Russian actors are caught up in discussions and attempting to counter pro-Ukrainian memes and comments, taking away time from creating and circulating propaganda and dis-and-misinformation online. Moreover, the pro-Ukrainan memes and comments ridiculing Russia cases raise concerns about the validity and seriousness of the attacked account (see Chapter 4).

Speeches, Symbols, and Images

Language, behaviour, and symbols are instrumental in creating a narrative that brings people together and allows them to communicate and develop networks. Memes can be a powerful tool in challenging dominant narratives, spreading awareness of underreported issues, and creating a sense of community among users who share similar opinions. Using symbols and visuals in memes can help quickly identify political realities and events and create a link between the viewer's emotions and political stance. To develop an effective memetic warfare strategy, there needs to be a coordinated communication approach that maximises the effect and maintains relevance and consistency in messaging. National symbols and images are essential tools in shaping public opinion, mobilising support, and creating a sense of unity and identity (see Chapter 5) (Knott, 2016; Miller-Idriss, 2016). Using symbols and images, such as the Ukrainian flag, can impact the audience nationally and internationally, while symbols associated with the aggressor can evoke negative emotions and be associated with the out-group (Denisova, 2020, p. 31). Therefore, using national symbols and images in the resistance against Russia can be a powerful tool in countering propaganda and promoting a national agenda. Using symbols, such as flags, in online profiles can help identify users and create a sense of belonging to a particular group or community. Different online groups have adopted this approach in countering propaganda, particularly against Russia, known for producing and circulating its propaganda and disinformation widely (see Chapter 3).

Humour, Irony, and Satire

Humour is a common element in many memes and can be used to engage the audience and convey a message in a lighthearted way. Memes that use humour can effectively challenge dominant narratives, provide alternative viewpoints, and spread awareness of underreported issues. It can also enhance the feeling of belonging to a community/having a shared identity. Humour can be a way to bring people together and create a shared sense of understanding and empathy (Taecharungroj & Nueangjamnong, 2015; Lindgren, 2022, pp. 33–34, 128–129).

Using humour, irony, and sarcasm in memes can help to enhance resilience among people who are affected by the Russian war against Ukraine, both the actions on the ground, but also in the online environment, where Russia is constantly communicating, planting false flags, and progressing the propaganda playbook known from previous conflicts and wars (see Chapter 3). Information and communication are essential to inform the population and supporters and counter falsehoods in a fight or conflict. Humour and satire in memes can help to subvert the propaganda and disinformation often used by opposing forces. By using memes and online architecture to promote a particular narrative, it is possible to create a counter-narrative that challenges the dominant rhetoric put forward by the opposing side (see Chapter 5) (Keblusek et al., 2017, p. 2; Denisova, 2020).

Challenges to the Use of Memes

The use of social media and digital environments is crucial for generating support and mobilising individuals, with online communities playing a significant role in connecting people and sharing information and resources. The book highlights the importance of incorporating memes and social media platforms in cyber defence strategies for supporting governmental and military actions (see Chapters 3, 4, and 5). The book also recognises that challenges are associated with this, such as the need for flexibility and adaptability to keep up with new technologies, behaviours, actors, and communication methods used in memetic warfare. The framework to manage memetic actions requires innovative and adaptable strategies incorporating public and private actors to effectively counter the constant flow of propaganda and dis-and-misinformation that is a part of hybrid warfare.

Memes have become a powerful tool for political expression, and their impact on public opinion should not be underestimated. Using memes and social media in warfare also presents challenges. For example, it becomes easier to spread false information and create echo chambers where individuals are only exposed to information that aligns with their beliefs and opinions. Therefore, developing innovative and adaptable strategies that effectively incorporate public and private actors to counter these challenges is essential.

It is possible that some countries or entities, including Russia, are collecting information about memes and their creators used by the defensive actors. This would not be surprising, given the potential impact of memes on public opinion. However, the effectiveness of such information-gathering efforts may be limited by the ability of memesters to protect their identities online. If a national strategy for memetic warfare were to be developed, it would be essential to consider the private actors' online security and privacy. Memesters and other online activists would need to be provided with the tools and knowledge necessary to protect their identities and communications. At the same time, it would be essential to ensure that such efforts do not infringe on the rights of individuals to express themselves online.

Box 6.1   Strategic Summary of Memes

Strategic summary

- The memetic approach adopted by Ukraine is defensive; Russia's model is offensive
- Social media and memes are effective in raising awareness, generating support, and promoting unity and resilience among people
- Potential for memes to spread false information and propaganda to impact on public opinion and potential threats to democratic processes
- A strategy requires considering the online security and privacy of private actors such as memesters and online activists:
  - Providing them with the necessary tools and knowledge to protect their identities and communications
  - Ensuring the protection of individual rights and freedom of expression

Public Actors – Communication Strategy

The use of social media as a means of communication and propaganda during military conflicts is not uncommon, and the Ukrainian government and armed forces have successfully embraced this approach. They have developed a communication strategy that includes coordinated online posts, speeches, and videos to reach a broad domestic and international audience. This strategy has effectively developed decentralised and fragmented meme forces online, with many groups and individuals engaged in memetic warfare on multiple levels, including creating and circulating memes, debunking propaganda and disinformation, and supporting the armed forces and the civil population.

The Ukrainian government's active presence on social media, including Twitter, Telegram, Facebook, and Instagram, has enabled officials and agencies to communicate with the public, share news and updates, and promote government policies and initiatives. The incorporation of coordinated governmental narratives should be central to the framework of future memetic warfare strategies, with communication between civil and military parts of the state apparatus being crucial in building resilience, as incorporated in the ROC's strategic framework (Fiala, 2019, 2020; Fiala & Pettersson, 2020; Liebermann, 2022). Overall, the success of the Ukrainian government's use of social media in communicating with citizens and the international community should be studied and incorporated into future memetic warfare strategies (Liebermann, 2022).

Social media have become essential for the Ukrainian government to communicate with its citizens and the international community. The lessons learned from observing this communication strategy should enable actors and agencies to develop a comprehensive framework that includes all communication forms, languages, symbols, and behaviors. To progress in developing a strategic approach, coordinated governmental narratives should be placed centrally in the framework. Contacts should be coordinated between the civil and military parts of the state apparatus as a part of the Total Defence approach and in terms of building resilience, as incorporated in ROC's strategic framework (Liebermann, 2022).

Challenges to Developing a Memetic War Strategy

Memetic warfare refers to using information and propaganda to influence public opinion and attitudes, and it has become increasingly prevalent in modern warfare. Skilled and non-skilled actors may work alongside national and international actors and agencies in this type of warfare, blurring the lines between civilian and military operations. Using civic actors to support military and governmental actors raises ethical and legal questions, as it can be challenging to define the roles and responsibilities of those involved in memetic warfare (Manjikian, 2021, pp. 291–292). For example, is it ethical for civilians to spread propaganda and dis-and-misinformation on behalf of a military or government agency? How is it possible to ensure that these individuals are not violating laws or regulations, for example, those related to privacy and free speech?

One of the critical challenges in memetic warfare is the overlap between civilian and military operations in the online environment. It can be challenging to distinguish between propaganda campaigns carried out by state actors and those carried out by independent actors, making it difficult to hold individuals or organisations accountable for their actions. To address these challenges, it is essential to establish clear guidelines and regulations for memetic warfare at the national and international levels. This could involve developing new laws or regulations that govern the use of information and propaganda in warfare and establishing mechanisms for monitoring and enforcing these rules. For example, the Tallinn manual recognises that civilians can participate in cyber actions along with armed forces and governmental actors; however, by doing so, they lose their protection as civilians, i.e. their classification changes to that of combatants (Manjikian, 2021, p. 294).

It is also worth considering whether it would be more beneficial to keep the civil actors in a decentralised network, where they act independently without interference by the state, to keep a distinction between the actors included in memetic warfare. Doing so makes it possible to circumvent the blurred lines between the groupings and prevent the state actors from setting up barriers for what the private memesters can do online and what narrative to pursue. Keeping the information flowing has proved useful, and the private actors decide what they want to engage with.

Box 6.2   Strategic Summary of Public Actors

Strategic summary

- The Ukrainian government and armed forces have successfully used social media to communicate with the public and promote their message
- Memetic warfare to counter propaganda and dis-and-misinformation, influence public opinion, and inform about current issues and events
- Challenging to distinguish between state and non-state actors in the online environment
- Clear guidelines and regulations to govern information and propaganda in warfare.
- Ensure that memetic warfare is used responsibly and ethically:
  - Ethical and legal questions about the use of civilians in memetic warfare should be addressed
  - Policymakers, military leaders, and civic actors work together to develop effective public and private strategies for civic engagement in memetic war

Private Actors – Civic Resistance

The phenomenon of memetic warf has become complex and involves various actors, including military combatants, civilians, private actors, and online
activists. The battleground for memetic warfare has shifted to social media,
where online communities mobilise to generate support for various causes
and resist online interference, propaganda, and disinformation. The spectrum
of actions involved in memetic fighting includes war and warfare, civil disobedience, activism, and individual strikes or campaigns. Notably, a shared
understanding, values, and beliefs bring these actors together in a public and
private, internal and external context.

The memetic concept includes civic resistance, which refers to non-violent
methods through fragmented networks of actors and groups that can successfully challenge the opposing force and undermine its objectives. The passage
suggests that to develop a successful strategy; certain elements should be enhanced, such as building a strong supportive community where individuals
can protect and support each other. However, there also needs to be a legiti-
mate cause of action and a level of communication between the government,
armed forces, and civic communities to progress a shared narrative, incitement, and activities on the tactical and operational levels.

Overall, the civic resistance and the use of ROC highlight the importance
of understanding the various actors involved in memetic warfare, the complexity of their motivations, and the need for coordinated communication and
action to progress a shared narrative and strategy. The passage also emphasises the potential of non-violent civic resistance as a powerful tool for challenging the opposing force in memetic warfare.

Voluntary Actors

In the context of ROC, civic resistance can be a powerful tool for resistance
forces to challenge the control and authority of Russia without resorting to
violence – and interfering with the operations of the Ukrainian Armed Forces.
By organising and mobilising the population and supporters of Ukraine to
engage in non-violent resistance, a large group of actors can work together
towards a common goal. However, to be successful in memetic warfare, it is
crucial to enhance the population's digital literacy in peacetime so that they
can use these skills in conflict to generate popular support for their cause and
counter propaganda and disinformation online.

Although these networks are developed outside the governmental framework, it is essential to have ties with governmental actors to ensure clear
communication routes online. In this decentralised format, civic actors can
pick up key speeches, arguments, or information from the government and
incorporate them into memes that can be circulated and reused in different
contexts. Furthermore, the passage highlights the different types of civic actor

involved in the resistance structure, such as hacktivist groups, meme forces, online groups, and individuals. Recognising and giving them the authority to continue their efforts is crucial.

The NAFO collective is also mentioned as an example of a group that empowers online users and produces a powerful tool of resistance against Russian propaganda (North Atlantic Fella Organization, 2022; Ukrainer, 2022). The open structure of the group, with no hierarchical leader group, makes it available for new members and provides flexibility to adjust to new situations. The solidarity among the members ensures that the memes are spread widely globally, enabling them to take control of the conversation and disrupt the Russian narrative while also boosting morale and strengthening their collective identity.

Challenges to Developing a Memetic War Strategy

There are limitations and potential drawbacks to relying solely on civic resistance in the context of the ROC conflict (Fiala, 2020). Non-violent resistance can be challenging to sustain over the long term and may be ineffective against highly repressive or authoritarian regimes. Furthermore, organising and mobilising a population for non-violent resistance can be challenging, and there needs to be a strong motivation for activating external online users and keeping them active for an extended period.

It is essential to consider the future of NAFO after the war ends and whether it will continue to exist or transition its power to other non-violent actions (Chenoweth, 2021, p. 240). NAFO is praised for its effectiveness, but its independence outside governmental and institutional structures can also be a weakness, as its effectiveness relies on its shared value system and synergy with the Ukrainian government (Chapter 4). Therefore, civic resistance should be used strategically and in conjunction with other tactics and strategies to achieve desired objectives. Organising these networks could potentially damage their freedom to target effectively. However, government-linked memetic units could support keeping the Fellas engaged in the memetic war and help defend Ukraine's interests online without introducing any form of censorship or control of the narrative. Access to defensive memes similar to those provided by the Ukrainian Memes Forces could be helpful for some actors to ensure a constant flow of memes.

Box 6.3   Strategic Summary of Private Actors

Strategic summary

- Challenges in developing a civic defence strategy with several actors; cooperation and exchange of knowledge should be encouraged

- Consider if the framework of the civic resistance should be formed within a governmental or non-governmental framework
- Upholding a sustaining non-violent resistance over the long term and the potential limitations of relying solely on civic resistance
- Potential benefits of government-linked memetic units supporting non-violent resistance efforts
- Civic resistance should be used strategically and in conjunction with other tactics and strategies to achieve desired objectives

References

CCDCOE, 2018. *Cyber security strategy of Ukraine.* [Online] Available at: https://ccd-coe.org/uploads/2018/10/NationalCyberSecurityStrategy_ Ukraine.pdf [Accessed 04 01 2022].

Chenoweth, E., 2021. *Civil Resistance. What Everyone Needs to Know.* Oxford: Oxford Univeristy Press.

CoE, 2017. *Doctrine of Information Security Ukraine.* [Online] Available at: https://rm.coe.int/doctrine-of-information-security-of-ukraine-developments-in-member-sta/168073e052 [Accessed 18 01 2023].

Denisova, A., 2020. *Internet Memes and Society. Social, Cultural and Political Context.* London: Routledge.

Fiala, O., 2019. *Resistance Operating Concept.* Stockholm: Special Operations Command Europe (SOCEUR) & the Swedish Defence University.

Fiala, O.C., 2020. *ROC. Resistance Operating Concept.* MacDill Air Force Base(Florida): JSOU Press.

Fiala, O. & Pettersson, U., 2020. ROC(K) Solid Preparedness. *PRISM*, 8(4), pp. 1–12.

Gurevitch, M., Coleman, S., & Blumler, J.G., 2009. Political Communication – Old and New Media Relationships. *ANNALS of the American Academy of Political and Social Science*, 625(1), pp. 164–181.

Hancock, B., 2010. Memetic Warfare: The future of war. *Military Intelligence Professional Bulletin*, 02 04, 36(2), pp. 41–47.

Iloh, C., 2021. Do It for the Culture: The case for memes in qualitative research. *International Journal of Qualittive Methods*, 20, pp. 1–20.

Keblusek, L., Giles, H., & Maass, A., 2017. Communication and Group Life: How language and symbols shape intergroup relations. *Group Processes and Intergroup Relations*, 20(5), pp. 1–12.

Knott, E. 2016. *Everyday nationalism.* [Online] Available at: https://stateofnationalism.eu/ article/everyday-nationalism/ [Accessed 15 03 2023].

Liebermann, O., 2022. *How Ukraine is using resistance warfare developed by the US to fight back against Russia.* [Online] Available at: https://edition.cnn.com/2022/08/27/politics/russia-ukraine-resistance-warfare/index.html [Accessed 24 12 2022].

Lindgren, S., 2022. *Digital Media and Society.* 2nd ed. London: Sage.

Manjikian, M., 2021. *Introduction to Cyber Politics and Policy.* 1st ed. London: Sage.

Mazarr, M., Bauer, R., Casey, A., Heintz, S., & Matthews, L.J. 2019. *The Emerging Risk of Virtual Societal Warfare,* Santa Monica: Rand Corporation.

Miller-Idriss, C., 2016. *The emotional attachment of national symbols.* [Online] Available at: https://www.nytimes.com/roomfordebate/2016/09/01/americans-and-their-flag/the-emotional-attachment-of-national-symbols [Accessed 24 03 2023].

Munk, T., 2018. Policing Virtual Spaces: Public and private online challenges in a legal perspective. In: *Comparative Policing from a Legal Perspective.* Cheltenham: EE publising, pp. 228–254.

Munk, T., 2022. *The Rise of Politically Motivated Cyber Attacks. Actors, Attacks and Cybersecurity.* London: Routledge.

Neudert, L.N., 2018. Germany: A cautionary tale. In: *Computational Propaganda: Political parties, politicians, and political manipulation on social media.* Oxford: Oxford Univeristy Press, pp. 153–184.

North Atlantic Fella Organization, 2022. *@Official_NAFO.* [Online] Available at: https://twitter.com/Official_NAFO/status/1577269391997050880?ref_src=twsrc%5Etfw%7Ctwcamp%5Etweetembed%7Ctwterm%5E1577274895095906305%7Ctwgr%5E60338d1736a091a06999d94fc562e0c9cc56eff2%7Ctwcon%5Es2_&ref_url=https%3A%2F%2Fnews.sky.com%2Fstory%2Fukraines-int [Accessed 27 12 2022].

Paul, C. & Matthews, M., 2016. *The Russian "firehose of falsehood" propaganda model.* [Online] Available at: https://www.rand.org/pubs/perspectives/PE198.html [Accessed 15 03 2023].

President of Ukraine, 2021. *Head of state approves strategic defense bulletin of Ukraine.* [Online] Available at: https://www.president.gov.ua/en/news/glava-derzhavi-zatverdiv-strategichnij-oboronnij-byuleten-uk-70713 [Accessed 24 12 2022].

Schmitt, M., 2013. *The Tallinn Manual on the International Law Applicable to Cyber Operations.* Cambridge: Cambridge University Press.

Schmitt, M., 2017. *Tallinn Manual 2.0 on the International Law Applicable to Cyber Operations.* Cambridge: Cambridge University Press.

Taecharungroj, V. & Nueangjamnong, P., 2015. Humour 2.0: Styles and types of humour and virality of memes on Facebook. *Journal of Creative Communications,* 10(3), pp. 288–302.

Ukrainer, 2022. *Who are the NAFO Fellas? The army of cartoon dogs fighting Russian propaganda.* [Online] Available at: https://ukrainer.net/nafo-fellas/ [Accessed 27 12 2022].

Zelensky, V., 2022. *A Message from Ukraine. Speeches, 2019–2022.* London: Hutchinson Heinemann.

Index

For Product Safety Concerns and Information please contact our EU
representative GPSR@taylorandfrancis.com
Taylor & Francis Verlag GmbH, Kaufingerstraße 24, 80331 München, Germany